Backyard Livestock

THE COUNTRYMAN PRESS

A division of W. W. Norton & Company

Independent Publishers Since 1923

Backyard Livestock

Fourth Edition

STEVEN THOMAS AND
GEORGE P. LOOBY, DVM

Illustrations by Mark Howell
and Patricia Witten

Raising Good,
Natural Food
for Your Family

COUNTRYMAN KNOW HOW

THE COUNTRYMAN PRESS

www.countrymanpress.com

A division of W. W. Norton & Company, Inc.
500 Fifth Avenue, New York, NY 10110
www.wwnorton.com

978-1-68268-086-5 (pbk.)

10 9 8 7 6 5 4 3 2 1

Photo Credits

Page 6: © georgeclerk/iStockphoto.com; 8: ©
HadelProductions/iStockphoto.com; 11: © vesmil/iStockphoto
.com; 14: © jackritw/iStockphoto.com; 26: © nikidavison/
iStockphoto.com; 30: © mira33/iStockphoto.com; 31:
© pets2and2/iStockphoto.com; 34: © cjp/iStockphoto
.com; 37: © chameleonseye/iStockphoto.com; 42: © ly-ly/
iStockphoto.com; 47: © vzmaze/iStockphoto.com; 54: ©
chuckcollier/iStockphoto.com; © georgeclerk/iStockphoto
.com; 59: © mansum008/iStockphoto.com; 62: © trangiap/
iStockphoto.com; 60: © bbbrrn/iStockphoto.com; 67: ©
Mike Stancombe/iStockphoto.com; 75: © Eremeychuk
Leonid/iStockphoto.com; 80: © Molotok743/iStockphoto
.com; 96: © flowersandclassicalmusic/iStockphoto.com; 108:
© BackyardProduction/iStockphoto.com; 113: © percds/
iStockphoto.com; 116: © JohnCarnemolla/iStockphoto.com;
118: © patrickjoseph1/iStockphoto.com; 131: © Sevaljevic/
iStockphoto.com; 146: © PavelKriuchkov/iStockphoto.com;
151: © brandtbolding/iStockphoto.com; 152: © Stieglitz/
iStockphoto.com; 161: © alinalina/iStockphoto.com; 166:
rachaelrussell/iStockphoto.com; 172: © Grigorev_Vladimir/
iStockphoto.com; 178: © lonewolfshome/iStockphoto.com;
192: © ngaga35/iStockphoto.com; 195: © 2ndLookGraphics/
iStockphoto.com; 197: © Hillview1/iStockphoto.com; 200 ©
narvikk/iStockphoto.com; 211: © ghornephoto/iStockphoto
.com; 213: © sonsam/iStockphoto.com; 220: © aon168/
iStockphoto.com; 225: © jesiotr9/iStockphoto.com

Discontent is the want of self-reliance.
—*Ralph Waldo Emerson*

For Meme, who fed the animals while
I wrote about them and without whom
none of this would be fun.

and

For Peter and Jane Jennison, for all
their help and encouragement over the
years and for their friendship.
—*Steven Thomas*

CONTENTS

Preface to the Fourth Edition

Change is inevitable. Sometimes it occurs at a leisurely pace, and sometimes it is frantic. In certain fields of endeavor it is almost impossible to stay current, the changes occurring with lightning speed. For those of us involved in any level of animal agriculture, change tends to take place at a more comfortable pace. But changes do occur. Almost ten years have passed since the last edition of *Backyard Livestock*, and changes have occurred, some for the better and some that remain to be evaluated.

In the years since the third edition was published, organic farming has grown in importance to become a major contributor to the nation's food supply. Many consumers have developed a strong preference for organically grown produce of all kinds. On a seasonal basis, farmers' markets have become an increasingly important source of food for those communities that they serve. The demand for land continues to be tight, especially in the northeast, and no longer is the housing market primarily driving the demand. Solar energy is now considered an important contributor. The producers of solar energy clear parcels of land on which to construct their facilities, posing an additional threat to prime agricultural land, of which there is a finite amount. It is time for the appropriate regulatory agencies to sit down and draw up a master plan that will address the needs of all concerned parties. Genetically modified organisms (GMOs) are also a major source of concern to many consumers, even though the leading scientists involved in their development assure us that there is no cause for alarm. Safe to say that the jury is still out on this one.

Global warming and all of its negative impacts must be monitored carefully, and those who have doubts about its existence need to study the evidence more carefully. It seems that many issues have been at least partially resolved, but Mother Nature is somewhat devious and we must be prepared for her next move—and we certainly must not contribute in any way to whatever tricks she may have up her sleeve.

As we worked our way through this revision, every attempt was made to retain as much of the original author's observations and recommendations as possible. Most of his comments are as true today as they were when written forty years ago. The good stuff doesn't change. When perusing this book, the reader will occasionally find the pronoun "we" when making a comment. That is Steve Thomas talking, not George Looby. Please feel free to correspond with us if you have corrections, omissions, or additions that you might want to address.

It is my strong hope that the material presented in this book proves to be worthwhile to you. Its purpose is to provide a quick and easy guide to some of the more common situations encountered in a modest start-up livestock operation. The material is based on the author's lifetime of experience in animal agriculture, some right, and some not so right—but if we can help you avoid even one pitfall along the way, we will have achieved our goal.

A Review of Current Topics in Animal Agriculture

The following section is designed to give the reader an overview of some the evolving issues that have surfaced since last we went to press. It is not designed to be an in-depth review of any of the topics, but rather to make the reader aware of their relevance in today's agricultural environment. You'll need to do further reading if you wish to expand your level of knowledge regarding any of these topics. It is likely that one or more of them would make another book!

GRAZING

Grazing is among the oldest and most efficient ways to convert grass into beef, and wool into textiles, but it is not always the easiest to accomplish. Land prices have soared as competition for land has intensified. Urban and suburban sprawl has made the ways in which land is used and managed very competitive. Small landholders who own modest farmsteads have been pressured by developers to sell their properties for inflated amounts in order to satisfy the apparently endless demand for housing in the suburbs and beyond. The tentacles of development reach out farther and farther, and the end is nowhere in sight.

The pressures cited above and others put increasing demand on land that previously might have been used for grazing. How long, you might ask, would it take for a few sheep or beef cattle to

give the same rate of return as a few acres sold to a developer for a large sum? The harsh reality is that, once gone, grazing land can never return, but the old adage about land being the ideal investment has never been truer that it is in today's real estate market. After all these years, Will Rogers is still right.

If, however, the real estate market has yet to peak in your area (and indeed I hope it has not) a modest grazing program might meet your needs. Pasture management can take many forms, so each manager must develop a strategy that best meets his or her own needs. Excellent pasture management is hardly a new concept, but it is now in the process of being rediscovered.

In the 1940s, the Greener Pastures program was developed in the northeastern states. Its goal it was to maximize the use of pasture for forage intake. The foundation of this program was pas-

ture rotation, in which the herd—usually the dairy herd—was allowed to graze on a parcel of pasture for a limited period of time, then moved to another parcel, then to another, which in turn allowed the previously grazed parcels to recover for the next cycle. The theory was that if the cow can be allowed to harvest the forage herself, production costs could be minimized. The downside of this management scheme is the considerable amount of forage wasted by cows trampling on it. Plus, when they excrete, the "meadow muffins" that accumulate promote the growth of grass, but this grass does not have the sort of flavor cows prefer. In your strolls through pastureland, you have no doubt seen clumps of lush green forage that are bypassed by the cattle. With aggressive management, this problem can be addressed by using simple machinery to spread the manure uniformly over the field. Harvesting by machine is a far more efficient method of harvesting, but only you, the manager, can decide which system is more effective for you.

To successfully manage this sort of grazing system, it is necessary to ensure that the grass is receiving the necessary nutrients to maximize its growth. Soil testing is a must. Anything less is guesswork. Most states have labs that perform this service for a nominal fee, and your county extension office can route you in the right direction to get this work done. Pastures in most sections of the country are limited to about 7 months—say, May through November—and even in the most temperate regions there are 2 months when very little pasture growth will take place. This means that additional feed must be available during those periods.

During the past few years, a group of beef cattle farmers have come to advocate a new system of grazing that they say will extend the grazing season well beyond its traditional limits. In the northeast, the end of the season is usually late October and sometimes into mid-November. The system involves taking the first cutting of hay off of a parcel in mid June, weather permitting, and then allowing the grass to grow undisturbed well into the fall, at which time the cattle are turned in. The advocates say that this cuts down on the time and expense of harvesting by letting the cattle do the work. In some instances they suggest that grazing may continue well into February, depending on the amount of land available and the number of cattle grazing. This can continue even though there is snow on the ground, as the cattle seem to have an instinctive ability to paw through to get to the grass underneath. Those with a traditional bent may view this with a bit of skepticism but it's instructive to consider the buffalo in the west wintering under the most severe of conditions.

THE MINOR BREEDS

Over the past two decades, a movement has developed to foster and preserve those breeds of livestock that have been relegated to minor status; breeds whose economic significance has diminished and whose future is in peril because of several interacting factors that need to be mentioned briefly in order that the reader be aware of their existence. During the early years of colonization, the settlers tended to bring with them animals that were readily available in their countries of origin. The majority of those settlers were of English origin, so the livestock brought to the colonies reflected that background. In truth, many of the distinctive breeds that we might recognize today were not well defined until the 19th century, when records of emerging types began to be carefully documented. An early breed type that was brought by the colonists would become the forebears of what we would recognize today as the Devon, a triple-use breed well utilized by those early settlers and for many generations thereafter. Devon cattle were particularly adapted for use as draft animals, as well as for milk and beef. Their lineal descendants are favorite draft animals to this day, and are often seen competing in pulling contests at fairs throughout the country, especially in the Northeast.

There are many reasons for the decline of certain breeds, but most often this was due to their failure to meet the criteria that other breeds were able to achieve in terms of milk production, rate of gain, carcass quality, and other measures of producer and consumer satisfaction. Those breeds that—despite many other commendable traits—could not meet the ever-increasing, stringent demands of the marketplace simply fell out of favor with the processors. Fortunately, many of what were to become these lesser breeds had a small core of advocates who were able to keep enough of their favorites alive that all their germ plasm was not lost. As mentioned elsewhere, today's processing plants demand an extremely uniform type of carcass for processing. Deviations from those requirements are neither encouraged nor tolerated, and as a producer you must conform or lose your market. For the small producer, meeting these criteria is not easy, so innovative strategies must be developed to meet the processors' needs.

The failure of many breed associations to adapt to the changing needs of the marketing system resulted in their near demise. One example of a breed organization that *has* adapted well to change is the Holstein-Friesian Association. The Holstein is by far the leading dairy breed in the country, outstripping all others by a wide margin in terms of production. It is estimated that 75 to 80 percent of all dairy cows in the country today are Holsteins.

Other dairy breeds may not measure up to the Holstein in terms of production, but they have other characteristics that are of value, especially when viewed over time. Each breed has unique characteristics that should be preserved by whatever means possible, thus ensuring that those characteristics are not lost. Extinction is forever.

Few of us are blessed with the foresight that would enable us to peer down the tunnel of time and predict what the needs of the food industry will be years down the road, nor can we predict what grave genetic mishap might befall any given breed or species. In the event of such a possible scenario, backup genetic material would be needed to fill such a gap, be it an act of nature or—even more likely—a human-generated gap. Too often we tend to become

far too narrowly focused, losing sight of characteristics that may seem trivial at the moment but have potential, at some point in time, to fill a real need.

It has been recognized that some breeds are better grazers than others, some are superior roughage converters, some are better heat-adapted, and others may possess unique disease resistance or parasite resistance. Failing to preserve these characteristics can serve only to make us more dependent on an ever-narrowing genetic base. Breeding for production characteristics alone may prove to be a fatal flaw.

There are organizations that promote the preservation of so-called minor or threatened breeds, one of which is The Livestock Conservancy (PO Box 477, Pittsboro, North Carolina 27312; www .LivestockConservancy.org). This organization publishes a newsletter that provides a wealth of information to its members.

The United States Department of Agriculture (USDA) has an active preservation program whose main thrust has been the development of banks of frozen semen from threatened breeds, storing it for any future perceived need. The main focus of the SVF Foundation (www.svffoundation.org) of Newport, Rhode Island is the freezing of embryos and banking of them in a secured facility. Thawing these embryos and implanting them into recipients has great potential as another means of breed preservation. In this program, representatives of endangered breeds are brought into the Newport facility, quarantined for a designated period of time, and bred. The embryos are then harvested and evaluated, and the best of the best are frozen for retrieval at some indefinite time in the future. This operation is a collaborative effort between the SVF Foundation and the Cummings School of Veterinary Medicine at Tufts University.

These programs have merit, but the backbone of any preservation program is still individual producers raising their favorite endangered breeds, either for sale or for use as breeding stock in their own operations.

The small farm should be a vital part of the preservation process. Where large commercial operations are often focused on maximizing production, smaller ones possess unique qualities that fit neatly into the minor breeds preservation programs. These smaller farms are more focused on developing niche markets that meet specific customer needs, such as specialty cheeses, unique sausage products, specific wool and fiber requirements, and an array of organically grown products. Breeding stock can be sold to enthusiasts who have developed a strong bond to one or more of the endangered breeds.

Among the cattle breeds that fit neatly into the small operation and are considered to be either on the critical or threatened list include the Milking Devon, the Canadienne, the Randall, the Dutch Belted, and the Kerry. Sheep breeds include, but are not restricted to, Gulf Coast, Hog Island, Santa Cruz, Jacob, Karakul, and St. Croix. Endangered pig breeds include the Hereford, Tamworth, Red Wattle, Mulefoot, Gloucestershire Old Spots, and Large Black. For further information on any of these breeds, contact ALBC (www.albc-usa.org/).

One downside is that marketing these animals through a nearby slaughterhouse can pose a definite logistical problem, as small independent slaughterhouses are disappearing. Without outlets located within a reasonable trucking distance that can accommodate a variety of different species, those animals that are of market weight and size, one critical part of any livestock operation may be compromised. You may be able to find local slaughterhouses by networking with other local livestock raisers, and check at feed stores, livestock auctions, and in the local or online Yellow Pages. Many backyard livestock raisers have resorted to selling animals on the hoof.

SUSTAINABLE AGRICULTURE

It is difficult to find a clear, concise definition of the term "sustainable agriculture." We have a good idea what the concept encompasses, but to state this precisely in a few sentences is not easy. That said, there is a great likelihood that if you are reading this book, you have already embraced at least some of the fundamentals of sustainable agriculture. When asked how they manage their farms, many farmers will come forth with the idea that they are merely stewards of the land, which they manage well while it is under their stewardship, and that they hope to pass on to succeeding generations a farm that is in better condition than when they inherited it. What is involved are good, sound, fundamental agricultural management practices that have been used in many operations for generations. Among these practices are good land management that conserves water, prevents water runoff, controls pollution, fosters biodiversity, and provides a boost to small rural communities, which may have lost their focus and sense of direction, due in large measure to the disappearance of small family farms.

Who then supports the concept of sustainable agriculture? The USDA funds a program called Appropriate Technology Transfer for Rural Areas (ATTRA), whose role is to provide information and technical assistance to anyone who is involved in sustainable agriculture at any level. The list includes farmers, ranchers, extension agents, and educators. The ATTRA Web site (www.attra.org) defines sustainable agriculture as environmentally sound farming that conserves and enriches the soil, protects the quality of water, and encourages a diversity of plant and animal species. It is not a prescribed set of universal practices, but rather a set of practices that are correct for a given farm, crop, or region. Among the systems that fall within the scope of sustainable agriculture are low-input agriculture, organic farming, biodynamic farming, and regenerative agriculture.

Interestingly, this program is not administered directly by the Department of Agriculture, but rather by an organization called the National Center for Appropriate Technology (NCAT). NCAT (www.ncat.org) is a private, nonprofit organization that uses federal funds to promote self-reliance through the wise use of appropriate technology. This program is directed mainly at low-income families.

ATTRA and NCAT cover many areas that readers of this book might find helpful, including such topics as field crops and soils, organic farming, livestock management, and education. Each falls under the mantle of sustainable agriculture. It should be noted that at this time, staffing levels do not permit easy access to the support staff for hobby farmers or students, but certainly if you have progressed beyond that point, I would suggest giving them a try at www.attra.org/who.html.

ORGANIC FARMING

Increasingly, we hear about organic farming and yet there seem to be more than a few misunderstandings about what it is or is not. Because small farmers often have a strong interest in pursuing their goals in as natural a manner as possible, it seems important to include a few comments about organic farming. Beyond that, the potential exists for supplemental income if you invest some time and investigate niche markets in your area. There have been many instances where very modest operations have evolved into something well beyond the original expectations of the operator.

Organic farming is a subset of the wider, more inclusive system called sustainable agriculture, as discussed earlier. It is a huge emerging segment of agriculture that would make a book in and of itself, and one that bears watching and careful study. If you are reading this, it probably applies to you.

The concept of organic farming is not a new one, but rather one that has resurfaced in the past decade to become economically significant. A subculture of consumers has emerged who not only favor organically grown produce, but insist on organically grown produce, meat, eggs, and dairy products. To purchase this type of product, they are willing to pay a premium price at the checkout counter.

There are many definitions for organic farming, but the basic premise suggests that organic farmers must develop systems to reduce pest organisms that are disruptive to production systems of any kind without the use of synthetic insecticides, fungicides, or other commercially produced products. Further, organic production systems are not allowed to use commercial fertilizers of any kind. A major component of any true organic program is that all steps in any production program must be carefully documented—a trail of paper must follow every step of the way. First a plan must be formulated, from inception to final sale, that outlines in detail the steps to be taken in implementing an organic program, be it squash, beef, avocados, or any other crop. Also, buffer zones must be established to ensure that adjoining or nearby nonorganic systems do not contaminate or otherwise compromise any organic system in place.

To what guidelines does the organic producer adhere? Organic farmers are not allowed to use synthetic pesticides or commercial fertilizers, and barriers must be developed to ensure that cross-contamination does not occur between adjacent farms. The guidelines also include no irradiation or genetically engineered foods or ingredients. As genetically modified foods come increasingly into the marketplace, the potential for a conflict between two diametrically opposed groups exists. The likelihood is that consumer groups will, over time, pick and choose those food and fiber sources that best meet their needs based on currently available information.

Who, then, ensures that all this works? As with any government program, which indeed this is, it was not necessarily designed to be simple! The ultimate authority for the implementation and management of the program falls under the authority of the Secretary of Agriculture of the United States, who then authorizes the Secretary of Agriculture in each state to assume the responsibility of managing the organic program. Interestingly, the various secretaries do not have their own agents monitor the program, but rather hire outside agents who are well versed in the mechanics of the program. These agents then make the actual on-site visits to ensure that individual farmers are carrying out the many aspects of the program.

If you have been considering hanging a sign on the maple tree out in front of your place with "organically grown" on it, perhaps it's best that you think again. Fines of up to $10,000 may be levied against anyone who misrepresents himself as an organic producer without following the mandates of the program.

Livestock producers who wish to be certified must sign up in a specific way and satisfy the mandates of the program. Prospective producers should contact their state's Department of Agriculture to be put in contact with an agent who will act to ensure that the applicant follows all program requirements. There is one advantage to being a small operation: A provision of the act states that farmers producing less than $5,000 worth of goods in a year are exempt from the provisions of the act. For anyone who exceeds that lower limit of annual sales and wishes to participate as a livestock producer, there are certain criteria that must be met. Record keeping is a critical part of the program. In the case of animals, everything that happens to them during their lives must be recorded, and the agent reviews those records on a regular basis. Buffer zones must be established between the organic producer and adjoining operations that are not organic to prevent possible contamination. Feedstuffs must be organically grown, and certain feeds are prohibited, including plastic pellets for roughage, manure refeeding, and feedstuffs containing urea. Growth hormones and stimulants of any kind are prohibited, as are antibiotics and synthetic trace minerals used as growth stimulants. The use of synthetic internal parasiticides

is not allowed. Vaccinations are allowed. The basis for disease and parasite control in organic programs is based on pasture rotation, balanced diet, strict sanitation, and control of stress. All these management schemes are based on the best available information, but they do have their limitations. Not everyone has the resources of time, land, and available feedstuffs to meet the required guidelines, but compliance is carefully monitored for certified organic producers. Given the scrutiny and oversight that organic operations must contend with, it is a given that they must command higher prices for their products than their conventional mass-production counterparts.

The concept of organic agriculture fits the goals of the backyard producer very well. The scope of their operations is small by most commercial standards, so they have the opportunity to ensure that the details of the program are well monitored. With greater individual animal contact, the mandates of the organic production programs should be more manageable.

ANIMAL IDENTIFICATION

The question of how to best implement animal identification has been debated for many years. One of the primary justifications for a good system of animal identification is traceability, i.e. being able to trace an animal or group of animals back to their farm of origin, especially when regulatory personnel are attempting to determine the origin of a disease outbreak. Much of the impetus stems from the outbreak of so-called "mad cow" disease (correctly called Bovine Spongiform Encephalopathy) in the UK several years ago, which was shown to be responsible for a debilitating disease in humans called Creutzfeldt-Jakob disease. If a given animal arrives at a destination showing evidence of disease, the route that animal has taken can quickly be determined via their identification, possibly averting a major disease outbreak. According to proposed legislation, all animals or groups of animals in the United States were to have been identified through a sophisticated computerized tracing network by January 2009. This proposed program was met with less than a high level of enthusiasm by the livestock owners of the country. Ultimately the program was dropped and replaced with a "voluntary" Animal Disease Traceability Program, requiring an official ID for animals moving in interstate commerce. Some states have instituted additional requirements beyond those of the USDA, such as certain age limits at which animals must be identified. This alternative program went into effect on March 11, 2013. It's probable that, no matter how small the operation, at some point in time animals will be either purchased or sold and interstate movement will be involved. In order to stay out of trouble with regulatory officials, new owners should be aware that regulations do exist and that state and federal regulatory personnel are ready to assist with any questions that may arise.

MARKETING

If the scope of your livestock operation is confined to supplying the needs of your family and perhaps some of your relatives, your marketing scheme is as simple as one can expect to achieve. You grow it and your brother-in-law and his family eat it—neat, simple, and to the point. No middlemen to dicker with.

Many hobbies or pastimes have a way of evolving into something far larger. And how does the hobbyist cope with unexpected growth? Better yet, do you have a plan that will help you identify a niche market, however small, to satisfy an unmet need in your extended community? If there is large Hispanic population in your area, you may find that you can capitalize on this market with a small capital outlay. Goats, for example, are easy to raise, and their housing and feed requirements are modest. The time from birth to market weight is short, and land requirements are minimal and quite basic. In most American households, goat meat does not rank high on the daily menu, amounting to less than 20 ounces per capita per year. This is not the case in the Hispanic market. Goat meat (chevon) is sold in a variety of different ways, ranging from whole baby goats to marinated cuts ideally suited for the barbecue. The Hispanic market prefers fresh meat products, not the prepackaged sort that so many customers have become accustomed to. The sale of even a very small number of kids can do much to offset the costs associated with any livestock operation.

Another product well suited to the backyard operation is designer eggs. More and more consumers are switching to eggs produced by cage-free chickens and organic eggs from hens that have never consumed products treated with commercial fertilizers or pesticides. The premium paid for these eggs makes it well worthwhile exploring the feasibility of marketing some of your surplus.

Another ethnic group that is emerging as a market to be served, and that particularly fits the capabilities of the small producer, is the Muslim community. The tenets of this religion mandate very strict rules regarding what may and may not be eaten and how it is to be prepared. For the small producer, meeting the code of *halal* should be considered when exploring marketing possibilities that fill the needs of Muslims. In Arabic, *halal* means "permitted," and that which is permitted is well-defined. A devout Muslim must do the slaughtering, and all blood must be drained from the animal. Pork is forbidden.

"Value-added" is something of a buzzword in marketing today, and any number of value-added agricultural products might be a viable sideline for anyone with a knack for processing them. Cheeses—particularly those that might fit into the gourmet category—find a ready market, especially in metropolitan areas. Sheep cheese has found a ready market, and for sheep raisers looking for new markets, this is certainly an area to be explored. Sausage making is an area that should be investigated if you find yourself with a surplus of pork and are lacking a ready market. Even your brother-in-law can only eat so much of a good thing.

Farmers' markets have become an increasingly popular outlet for locally grown products of all kinds, including meat, cheese, eggs, and even soap. These markets are usually conducted once a week at a convenient location in a community, such as a town green or common. Here, area farmers can set up modest outlets, such as from the back of a pickup truck, to sell to con-

sumers looking for fresh-from-the-farm produce of all kinds. These markets are usually run by local individuals and overseen and promoted by state departments of agriculture. It should be noted here that the terms "fresh," "local," and "native" have no legal meaning and should be viewed accordingly. At these gatherings, backyard operators may often sell their surplus jointly with a larger, more established operator. These gatherings frequently turn out to be community get-togethers where local news and gossip can be freely traded while getting "fresh" goods for the family table.

BIOSECURITY

In light of the world situation today, biosecurity has been added to our everyday lexicon. Biosecurity might be defined, in part, as those measures that a livestock operation must take to ensure that malicious acts of violence are not allowed to take place on the secured premises. Isolation, quarantine, and vaccination have long been staples when establishing good husbandry practices, but in today's increasingly unstable world, more rigid regulations are called for to ensure that the health and viability of our flocks and herds is not compromised. The small operator may be inclined to believe that threats from the outside are of little concern. Unfortunately, the potential for risk knows no boundaries, and everyone is at risk. The recent foot-and-mouth disease outbreak in the UK points up the fact that if the security mechanisms are not in place at those farms that are at risk, the entire animal health system may break down. Currently, the specter of avian flu hangs like a Sword of Damocles over the poultry industry worldwide, with the added threat that the causative virus may mutate and precipitate a human pandemic. The potential for some radical group attempting to do great harm to the animal population is ever with us.

There are at least five areas that need to be monitored if there is to be a successful biosecurity system put into place, and this applies to all facilities where poultry and livestock are housed, regardless of size:

- Maintain existing management procedures
- Attempt to keep a closed herd or flock wherever possible
- Monitor temporary breakdowns in biosecurity due to inconsistent management procedures
- Manage groups of animals within a herd or flock with disease control and prevention as the top priority
- Quarantine new stock before introducing to the rest of the herd or flock.

LIVESTOCK PURCHASE

- Buy additions from as few farms as possible, and only from those with a known health status.
- Discuss with your veterinarian whether any laboratory tests should be done prior to purchase.
- Request all pertinent documents relating to the health of the animals to be purchased.
- Avoid buying at an auction unless the health status of all animals offered for sale is well documented.

- Use your own truck to transport purchases; avoid livestock haulers whose background is not well known to you.

QUARANTINE

- All purchased additions should be quarantined for six weeks before commingling with existing stock. Breeding stock may require a longer period of quarantine.
- Quarantine means complete isolation—additions can have no contact of any kind with the resident population. Care should be taken so that the resident pet population cannot move from the isolation facility to other housing on the same farm. Free-ranging poultry should be restricted in their movements. Accomplishing this in a small operation can present some logistical nightmares. Most of us do not have the facilities or acreage to allow us the luxury of complete isolation, so we do the best that we can.
- One individual should be assigned the sole responsibility of caring for any animals in quarantine.
- Animals should be examined on a regular basis, and the exam should be noted in a health notebook, signed and dated.

FARM SECURITY

- Secure and monitor all farm entry and exit points. Boundaries should be observed regularly and unusual activities carefully noted.

- Limit visitor access to livestock holding areas and avoid direct contact with animals. Keep unauthorized personnel from entering farm buildings.
- Take necessary precautions to ensure that authorized visitors take measures to maintain high levels of hygiene upon arrival and again upon departure. This would include contractors, veterinarians, inseminators, feed delivery people, and anyone with regular business on the premises. This would include but not be limited to boot cleaning both before and after the visit, clean outergarments, clean equipment, and disinfection of vehicles. Provide disposable plastic boots to all casual farm visitors.
- Anyone who fails to comply with your regulations should be denied access to the farm.
- When stock is away from the farm (at shows, fairs, etc.) take all necessary measures to avoid contact with any stock whose appearance is in any way suspicious.

HYGIENE AND THE FARM ENVIRONMENT
OUTSIDE
- Ensure an adequate supply of clean, fresh water at all times.
- Move feeders regularly to avoid a having a feeding area deep in mud.
- If possible, avoid spreading manure on pastureland.
- Use good pasture management practices.

ANIMAL HOUSING
- Provide adequate ventilation, but avoid drafts at the animals' level.
- Clean the housing facilities as frequently as possible to prevent the buildup of possible infectious agents, and especially feed bunks and waterers. Provide foot baths for visitors, the help, and for the stock.
- Where possible, manage animals in small groups. This provides a less stressful environment and allows for better observation of potential problems as they develop.

BARNYARDS AND FEED STORAGE BUILDINGS
- Keep trash, garbage, and debris to a minimum to keep the vermin population at the lowest possible level.
- Secure feed storage areas against vermin, birds, and cats, and ensure that they are also watertight.

As an aside, this writer has long felt that birds are a major vector or carrier of disease to our herds and flocks. How easy it is for a bird visiting a farm on which there is a disease outbreak, no matter of what magnitude, to feed in the bunk or feeder at which an ill critter is feeding, and then fly away to a neighboring farm and spread the virus, bacteria, or other pathogen to the animals in that operation and never suffer any ill effects itself.

DISEASE PREVENTION, SURVEILLANCE, AND RESPONSE TO PROBLEMS

- Prevention is the key to any good program. Reaction to a problem is never the best approach. Have a plan in place that addresses the issue of potential problems.
- Record keeping is important to identify problems while they are still small ones, spot trends that may be subtle but important, and ensure that things that should be done are indeed done. One good adage to remember is that if it isn't written down, it didn't happen. Further, a dull pencil is better than a sharp mind.
- Use laboratory backup to identify latent, subclinical situations that may be festering just below the surface yet impacting the overall production level.

Antibiotics in Animal Feed

On January 1, 2017, federal regulations took effect that drastically restricted the use of antibiotics in animal feed. For almost six decades, antibiotics have routinely been added to various animal feeds based on research that this supplementation enables animals to process their feed more efficiently, thus increasing the rate of weight gain on less feed. Further, the addition of antibiotics reduces the likelihood of low-grade intestinal infections to which younger animals are often prone. This practice has been under review for several years, with many experts in the field of human medicine speculating that the practice may have been contributing to the alarming rise in the number of cases involving antibiotic resistant bacteria seen by physicians in practice, most often in hospital settings. Animal scientists were not easily convinced that their use of antibiotics in animal feed could be contributing to the development of so-called super bugs, but the concerns of millions of concerned citizens across the country could not be ignored.

The regulatory process at the federal level does not always take place with lightning speed, and so it was with this issue. Hearings were held, the testimonies of panels of experts were received, impact studies were conducted, and, all the while, stories of the super bugs in hospitals throughout the country continued to surface. There are more than 25 pharmaceutical companies manufacturing antibiotics that supplied the huge amount of product that went into animal feed. In 2012, 19.6 million pounds of product was approved for animal use, and nearly all of it was sold over the counter with no prescriptions needed. When the FDA began its review process, it was determined that all pharmaceutical companies would have to refrain from the further production of antibiotics for over-the-counter sale. This was not an easy sell on the part of the agency, so it was forced to build a penalty clause into the program so that implementation could move forward.

With over-the-counter sales greatly reduced, livestock producers will be required to work more closely with their veterinarians as all antibiotics will be sold by prescription only. What this means is that only animals showing signs of an antibiotic responsive illness would be able to legally receive antibiotic therapy. For many veterinary practitioners, this new regulation will require a period of adjustment to one of far greater oversight in the way in which antibiotics are used on the farm. Compliance by producers will be critical to the success of this regulation. Given the number of producers to be affected, avoidance will be fairly easy.

Most of the changes now in effect will not impact the backyard livestock producer, but they should be aware of the issues on the national scene. Also, if they have been buying feed from a large mill, they should be certain that there are no longer antibiotics in the feed that they buy.

Animal Rights Issues

Increasingly, concerns about animal rights pose potential problems to producers of any size. Animal rights groups are sometimes somewhat misguided in their perception of what should and should not be allowed in production agriculture with regard to how animals are managed. Running free with no restrictions is often not the best way to manage animal groups, especially those on a parcel wedged between I-91 on one side and the Conrail freight line on the other. When dealing with animal rights agendas, it is probably best to try to avoid a confrontational approach. The issues involved are extremely emo-

tional, and all sides usually have rather strongly held views. The emotional side of animal rights issues often prevents them from being discussed in a calm, logical, orderly manner without one side or the other side becoming somewhat out of control, thus preventing any adequate understanding of the other side's views. There are no easy answers to some of these questions, but there should be an ongoing dialogue between the opposing factions with the hope that somewhere down the road each side may come to a better understanding of the other group's views.

When the conditions are just right, those who have trouble accepting practices involved in agriculture might be reminded that, if there were no need for animals for meat, milk, wool, or any other animal product, the need for those animals would disappear and they would be gone.

Cloning and Genetic Engineering

For many consumers, the idea of purchasing meat, dairy, and plant products that have been artificially altered in any way runs contrary to their basic instincts of what constitutes wholesome food. To suggest that a product has been genetically engineered conjures up in some minds images of a slightly demented scientist fiddling with Mother Nature in a manner that bodes no good for the consumer, or for that matter the entire human food chain. It is well beyond the scope of this little book to enter into the ongoing discussion on this topic, but there are a few points that might clarify some of the issues.

Genetic "engineering" is a relatively new term in livestock production, but in truth, while people have selected for certain characteristics since they first began to domesticate animals, genetic selection has been practiced intensively for the better part of 200 years, dating back to the time when breed registries first began to emerge. By then, a breeder could predict with some confidence that the offspring of a particular pair of animals would have the same physical characteristics as its antecedents.

An issue that bothers many people today is that the process has become much more sophisticated, with much of the selection being done at the cellular level rather than at the animal level. This sort of human intervention is a worrisome thing to many people, who look at the whole process as intruding on the natural scheme of things. To some degree this may be true, but the quest for continual upgrading and refinement is ongoing, one that is unlikely to go away, being as it is a component of the human spirit.

Genetic engineering is a step forward (though some might dispute this) in the selection process, perpetuating bloodlines of superior stock using cells of the animals themselves to reproduce the traits that have made them superior. Agriculture today is focused on preserving and intensifying those traits that made a particular animal within a breed particularly outstanding when compared to others within that same breed. Too much concentration of a partic-

ular bloodline within a given breed may have a significant downside. The possibility exists that while certain desirable traits are being intensified and preserved, other latent, undesirable characteristics might also be preserved to emerge at some point in the future to cause chaos in the breed in question.

Support for Young Farmers

For most of recorded history, young people exploring a career in farming had few if any resources to turn to for assistance in their decision-making process. Recently there has been a significant change as a number of organizations have developed programs to address the needs of this group. One of the factors that prompted this change has been the number of returning veterans who have expressed a strong interest in agriculture as their post-service career. Many of this group, both veterans and non-veterans, have chosen not to pursue additional schooling and instead go the hands-on route to get the necessary training and experience they need to be successful in one of the many agricultural fields open to them. Another factor that has prompted the several agencies to develop new programs is the aging farm population. There are many individuals approaching retirement age who have no children interested in following them but want very much to have their land to be continued to be farmed.

The USDA has taken a lead role in promoting and encouraging start-up farmers to make use of the financial resources that are available. The Extension Service of the University of Connecticut has a scaling-up program for new and beginning farmers, and it is likely that extension programs in other states are similarly involved. For those who are considering organic farming, the Connecticut Northeast Organic Farming Association has a variety of programs thoroughly explaining their organization and its benefits. The New Entry Sustainable Farming Project located in Lowell, Massachusetts was originally established to provide assistance to individuals newly arrived in this country who wanted to become established as farmers. The role of this program has expanded to include everyone who has farming as a career goal.

The departments of agriculture in several states have programs whose primary function is to lend support to this ever-growing segment of the population. Within the state of Connecticut, the Connecticut Farm Bureau is supportive of all things that relate to agriculture in any way, and so it is with its support of young farmers programs. The New Connecticut Farmers Alliance is, as the name implies, a group of young farmers who, among their many activities, lend support and guidance to those who wish to join them in a career in farming.

Why has this overview been included in this book? For those who may be considering the purchase of livestock of any kind for the first time, this plunge could be something of a slippery slope. It would not be the first time that an individual has purchased one animal as a curiosity, only to see it evolve into a full-time endeavor. So if you find that livestock management, with all of its ups and downs, becomes your primary focus, know that there is help ready and willing to lend a hand when and if the need arises.

Introduction

I think the best advice for people beginning to grow their own meat is to start slowly. Raising animals is a most enjoyable and satisfying experience, but it can be a nightmare if you get in over your head. We started a bit too fast. We had a couple of horses and some chickens, then we got a couple of pigs and some sheep, and before we knew it, the Vermont winter was almost upon us and we had no suitable shelter for our now-pregnant animals. We did, somehow, manage to get our animals settled for the winter (in the case of our pig, 2 days before she farrowed), but we vowed not to get any more animals until we had pens constructed and we fully realized how much additional time each animal would take. Table I.1 will give you a rough idea of how long each animal (in the case of poultry, a dozen birds; sheep, six to twelve head) will take of your time each day, summer and winter:

Table 1.1: Approximate Time per Day to Care for Livestock		
ANIMAL	**TIME/DAY (MINUTES)**	
	SUMMER	**WINTER**
Poultry	5	7
Sheep	5*	10
Milk goat	15–30	15–30
Pig	5	7
Rabbit	5	7
Veal calf	7	10
Beef calf	5	10

*If you raise grass lambs that are contained by fencing, you will have practically no time investment (except to check their general well-being) other than to provide water. If you grain them in the summer, figure an additional few minutes.

The difference in time required between summer and winter reflects the need to carry and thaw water and to clean pens (since animals will spend more time indoors in the winter months). The time you'll need to care for your stock will depend on how far you have to carry water, how well organized your pens are, etc. Don't let the amount of time scare you if you're figuring on keeping a number of different animals. If you have three pigs, a flock of chickens, and a dozen sheep, it probably won't take a full 7 minutes per pig, plus ten for the sheep and seven for the chickens in the winter, for a total of 38 minutes per day. There is a lot of overlap. A big pail can be filled up to supply water for all the stock; pigs are fed the same ration, near each other or in the same pens. Above all, with experience you will organize your barnyard, or yard, so that your time will be used most efficiently.

The animals I have included in the book were chosen for ease of care, for relatively small space requirements, desirability of products and, most importantly, their ability to forage for their own food or to make use of low-protein food, otherwise wasted foodstuffs, or the products of other animals.

The most important feature of this book, in my mind, is the section on *Supplementing Commercial Feed,* found under the *Feed* section. Here I have attempted to give you some independence from the feed companies by pointing out other feeding routines that you can implement yourself. Let's be pessimistic and assume there'll be grave problems feeding the world in coming years . . . or, if you will, let's just be practical: Feeding your animals without buying commercial feed is a cheap way to eat! This is just a beginning. I have collected some methods, either from my own or other people's experience, of alternative feeding systems. I should warn you, though, that I have not personally tried all the supplemental feeding routines, so use some caution when implementing them.

I have tried to organize this book, with its chapter divisions and subdivisions, to make it easy to refer to any aspect of livestock care. Most chapters are divided into nine major sections, listed below with a description of each:

1. **Breeds:** I have not tried to acquaint you with all major breeds within a species; however, the most common and useful for a backyard operation are discussed, descriptions are given, and in some cases illustrations are provided.

2. **Purchase:** The how and when are discussed, as well as the whys of purchasing a certain animal. Features to look for, and avoid, are offered for each animal. This also includes information on how to care for the young animal. In the case of poultry, hatching as well as brooding facilities and methods can be found. To get a firsthand idea of what a good specimen of a given species looks like, go to 4-H clubs, livestock shows, or county fairs and inspect prizewinners.

3. **Housing:** Simple shelters for summer, as well as more elaborate winter quarters, are described. Also included are details of fencing and other methods of confining your stock.

4. **Equipment:** Feed dishes, water systems, and other day-to-day needs are discussed and simple plans shown to enable you to build your own. More specialized equipment for breeding and health care are covered in their appropriate sections.

5. **Feed:** This section is divided into (a) Conventional Feed and (b) Supplementing Commercial Feed. The first subdivision covers those feeding practices using commercially purchased feed, hay, or the like. When buying your feed, be aware of the going prices. Feed stores in one area often charge sizable differences in prices for the identical or comparable feed. Feed is usually cheaper in summer. Whenever the prices takes a sharp drop, buy in quantity if you can afford it and can store it properly. The proper watering of stock cannot be stressed enough. Water should be available, thawed, at all times.

Supplementing commercial feed has been mentioned above. In addition, topics in this section include mixing your own feeds, free or very cheap sources of feed that can be substituted for commercial feed, and hints on how to stretch your feed dollar and make your meat cheaper. If you focus on collecting supplemental feed—and make a real commitment—you will be amply compensated by lower meat costs.

6. **Management:** This section has three subdivisions: (a) Routines, (b) Handling, and (c) Predators. It also contains information that does not fit in any other section.

"Routines" deals with the management of your animals. Suggestions are made as to when to raise particular animals, in what numbers, and how to establish the most efficient and money-saving regimen of raising an animal for your family.

"Handling" involves just that—the correct (and

easiest) ways to catch, transport, or carry particular animals.

"Predators" points out the more common members of the wildlife community that will prey on your stock and how to prevent or eradicate them. As a rule, a good barn cat will do a fine job protecting your investment in animals and feed from rats and mice.

Keeping accurate records is an important facet of any livestock raising operation. Those "bargain" eggs you think you're getting may not turn out to be such a great deal when you calculate how much feed is going into the production of each dozen eggs. It is easy to fool yourself if you don't keep records. You might remember only four bags of feed you bought for your pig when you actually used eight, so keep track. I like to keep a separate book on each animal (or flock, herd, whatever) to see how they pay off in the long run. Depending upon your aims—profit, break-even, or just good eating—you can make an accurate evaluation of how each animal is producing and decide on the

most efficient routines for your family. These records should also contain other pertinent information, such as breeding dates and outcomes, butchering weights, overall cost per pound, and the like. These records are also fun to look back on over the years to see how efficiently you've learned to run your operation. It's also nostalgic to look back at how cheap feed used to be. (Good records are also necessary if you are filing income taxes as a farmer.)

You will discover in time that the concept of backyard livestock makes for better stock. In raising animals, smaller numbers are best. Diseases are limited, for the most part, to animals that are overcrowded and lack proper sanitation, housing, and ventilation. You shouldn't have any such problem with a backyard flock. In larger operations, an owner simply doesn't have the time to check each animal daily and separate those animals that aren't competing well for food. In a small operation, you will have few, if any, low-grade animals because you can offer individual attention. If one of three pigs is being bullied you can, without much bother, make another pen for it or feed it at some distance from the other two. A large-scale owner must accept such poor competitors or take a financial loss and cull them. I can see it with our sheep. When we had a half-dozen, they all got their share. As our flock grew, and separate feedings and other "coddling" became impossible, "poor competitors" became apparent. I had to take measures to ensure they all got their share or resign myself to a few unthrifty sheep. In short, our animals are superior because I can watch each one

carefully and spot any bullying, signs of disease, or other problems early and correct them before there is any permanent damage. In a larger operation, this is far more difficult and time-consuming.

In managing your animals, try to spend some time with them every day. The first sign of disease is often a drop in feed consumption, and if you're attentive, you'll spot this right off. By standing and watching our animals as they run in the barnyard or graze, I can spot even the slightest abnormali-ties: foot problems, lameness, cuts, or other unusual behavior. Spend some time with them and, above all, *be nice to them.* They are furnishing you with a most precious product. Pet them, talk to them, scratch them—never be mean to them—and they will pay you back.

7. **Breeding:** This section has details on breeding practices. It includes information on when and how to breed, gestation, feeding during pregnancy, and

equipment for breeding and birthing, as well as weaning and castrating.

8. **Health:** The key to guaranteeing the health of your animals is prevention. This concept is reiterated so often throughout the book that you may get sick of it, but it does bear repeating. If you provide your stock with proper amounts of food and water, clean quarters, and fresh air and sunshine, you will—unless there's a local epidemic of an infectious disease—have a disease-free operation. This section lists the most common afflictions, such as parasitic infestations (worms), and their prevention and treatment. A disease table for each species, including major afflictions, their causes, symptoms, prevention, and treatment, is presented in Appendix B for easy reference. These charts should aid your quick diagnosis and treatment of major diseases. If you are really serious about raising livestock, you would do well to purchase one of the veterinary manuals listed in Appendix F.

9. **Butchering:** Complete butchering information is given for all poultry and rabbits. The procedures for the remaining animals are too detailed to be given adequate coverage in this book. If you watch them being butchered once or twice and use a book on butchering, you should be able to master the procedure yourself.

You might be able to sell or barter some of your surplus meat to further reduce your costs, but check state laws first. Many states require any meat for sale to be butchered only in state-registered slaughter houses. You do, however, have the additional option of selling the animal alive and having the purchaser arrange his own butchering.

The "Grow Your Own..." chapter is designed to be used in conjunction with the "Supplementing Commercial Feed" section and gives you an introduction to growing your own grain crops, pasture, and making hay and silages. There are also references in Appendix F that you can use to obtain additional, detailed information on growing your own feedstuffs.

I have attempted, within sane limits, to make this a one-book reference. I hope that it will suit both the person who wants to raise a few laying hens and a lamb and a pig, as well as those who wish to get more deeply involved, raising animals on a fairly large scale—arranging their own breeding, and doing their own castrating, butchering, and the like.

We are lucky to have livestock and poultry that consistently grow fast, live long, are disease-free, bear large and healthy litters, etc. If anyone were to ask me if there is any sure way of raising superior livestock, I'd have to say yes. Simply do it by the book (I'm not being immodest—you needn't use this book—do it by any book). Do what you're supposed to do. Furnish your stock with plenty of fresh water, a proper and clean pen, the correct amount of feed, and that's it. The stock I see having problems usually aren't getting water, or their owners are taking other shortcuts. Such people suppose that, for example, by cheating on feed a bit, the pig will grow anyway and be cheaper. Not so. If it says to feed your nursing pig ten pounds of feed a day, do it. Sure, she'll probably make it on seven pounds; maybe won't even lose any pigs, but she'll be run down and have a shorter productive life. Then see if you get a reputation for having good piglets!

Finally, I don't profess to be an expert. Not by a long shot. Since I have finished this book, I have learned countless things that I should have included. Many people know more about animals than I do and have raised them for more years, but I guess they didn't want to write about it. There are no "experts," you'll learn every day no matter how long you've raised animals. In that sense it's presumptuous to think that this book is complete. Refer to other books in the References, pump other people for information, keep your eyes and ears open, and you'll be surprised at how fast you'll learn. It may have taken you eight years to learn your multiplication tables, but you'll learn about farrowing the day your sow is due.

CHAPTER 1

Poultry

Chickens

Of all the animals that anyone could choose starting out in a backyard setting, the most suitable one is the chicken. The start-up costs are low, and the chicken's needs are modest. Your own time commitment should be low, too, unless you just enjoy watching hens scratching in the yard.

Since this book was first published, the role of eggs in our diet has been in a state of flux: first on the forbidden list and now enjoying almost a complete reversal. It appears that eggs are OK again.

Nutritionists increasingly advocate chicken as a major portion of the total meat intake in our diets. Raising chickens at home for this purpose is a most practical consideration.

BREEDS

Almost 200 different breeds of chickens are listed in the *American Standard of Perfection*. If you peruse a poultry catalog, you will be amazed at the many unusual and downright bizarre breeds

of chickens, but for our purposes we'll classify chickens into three groups (1) "egg" birds; (2) "meat" birds; and (3) "dual-purpose" birds (Table 1.1, Figure 1.1). The egg birds are small and are excellent egg producers, but because of their size they are rarely raised for table use. Meat types are large, broad-breasted birds that convert most of their feed to meat and are usually poor egg producers. Dual-purpose birds incorporate the qualities of both egg and meat birds into one and enable the homegrower to make an enjoyable meal out of any egg producer that has reached the end of her production years.

If you choose one of the egg-laying breeds, remember that the majority of breeders have so refined their egg-laying characteristics that these chickens will lay well for only one or two seasons.

Of course, your choice of breed will depend on the needs of your family. Once you taste rabbit, you may simplify selection by getting a few layers and relying on your rabbits for meat, but that's another consideration altogether. Among dual-purpose

Table 1.1: The Three Classes of Chickens and Common Breeds Within Each		
EGG BIRDS	**MEAT BIRDS**	**DUAL-PURPOSE**
Leghorn	Cornish	Plymouth Rock
Minorca	Orpington	Rhode Island Red
Ancona	Australorp	New Hampshire
Blue Andalucian	Brahma	Wyandotte
	Cochin	
	Langshan	

birds, I have found Plymouth Rocks superior to New Hampshires and Rhode Island Reds in meat quality, and about equal in egg production. The White Plymouth Rocks dress out more attractively than the Barred Plymouths, with no dark feather shanks left in the skin, but you can often sell the neck feathers and skin of the Barred Rock roosters to sports stores for use in fly tying. (To prepare the skin, take the skin off the neck with feathers intact, and salt well to preserve it.)

By experimenting and trying other kinds of birds, you will find out which are best suited to your needs and taste. Most commercial egg and meat birds are hybrids. Our first birds were bantam-Rhode Island Red crosses. They combine the docility and egg-laying ability of the bantam with the somewhat larger eggs and better meat quality of the Rhode Island Red. For the ambitious poultry raiser, cross-breeding opens up an interesting opportunity for experimentation. I remember a story in the news a few years ago about an Australian teenager who, by crossing at random different breeds of chickens, came up with a bird that matured to 26 pounds and laid one-pound eggs! Panic-stricken poultry companies offered a million dollars for the bird so they could *destroy* it and with it any competition for their rubbery chicken and mass-produced eggs. Since I have heard nothing else about it (and have yet to see 25-pound chickens in the supermarket), I can only assume the boy accepted their offer and now lives a life of leisure munching buckets full of Kentucky Fried Chicken.

PURCHASE

To decide on the number of chickens you'll need, take into account your egg and meat consumption and determine whether the birds are intended for home use or whether you plan to defray their feed costs by selling surplus eggs. As an aid in helping you figure out the number of birds you need for your particular situation, figure a good layer in her first year of production will give you 200 to 220 eggs. The first year of production is the best year, and in each succeeding year a hen produces fewer eggs. That decrease depends on health, breed, age, genetic makeup, and most of all, the care you

give her. Generally, you can depend on almost one egg per day per bird until she goes into molt (see *Management* section), when production usually ceases. As far as meat is concerned, again depending on breed, you can expect a bird to dress out at between 2 and 5 pounds at the age of 5 or 6 months. If you plan to market eggs in your neighborhood, you can buy as many as you have space for and can handle financially, but don't undertake more than you think you can market. However, if you are just starting out, it is best to fill your family needs and worry about selling eggs and/or meat at a later date.

Once you have decided on your breed and quantity, you are ready to make your purchase. You will have the option of buying mature birds (i.e., birds of production age or older), day-old chicks, or fertile eggs for hatching. Your initial plunge would prob-

FIG. 1.1: EXAMPLES OF COMMON BREEDS OF CHICKENS

ably be easiest if you purchase ready-to-lay pullets from a poultry farm. This frees you from the problems of hatching and brooding chicks, plus the wait of 4 to 6 months for your first eggs, and the discouragement of losing them at this young age due to disease or beginner's ignorance.

Mature Birds
If you take this option, you are ready to gather farm-fresh eggs the next day (or, as in our case, on the ride home). Often, depending on your source, it may be cheaper to buy mature birds rather than feeding them for the five months it takes for chicks to begin producing. You might purchase them from a nearby poultry farm or from an overstocked neighbor. Buying from a livestock auction is not a good option. How did they end up there in the first place—are they someone else's culls? Don't get stung on your first venture into the business; there will be plenty of opportunities for that later on.

The disadvantages of buying mature birds are that you will probably be limited in your choice of breeds, and if you're not careful you might buy poor-laying or nonlaying birds. Table 1.2 outlines characteristics to look for in choosing a good laying hen. This can be used both in purchasing your birds and in culling birds from your flock.

There is still a bit of folklore that suggests that you should have a rooster in your flock to maximize production. This is not the case. The only reasons for a rooster are to have a readily available source of fertile hatching eggs, or if the sight and sound of a rooster rounds out your idea of what your place ought to be. Recent news releases tell sordid stories of roosters introduced into suburban neighborhoods and their unhappy fate.

Day-Old Chicks
Buying just-hatched chicks enables you to have a wider choice of breeds, to watch them grow, and have that inexplicable joy of finding that first walnut-sized egg five months later. Chicks are usually sold in three classes: 95 percent pullets; as-hatched, or straight run (not sorted by sex); and cockerels.

Table 1.2: Characteristics of Good and Poor Layers		
CHARACTERISTIC	**GOOD LAYER**	**POOR (OR NON) LAYER**
General health and appearance	Alert, well proportioned, active	Deformed, listless, weak
Eyes	Bright and alert	Dull, listless
Comb and wattles	Thick, smooth, and bright red	Pale, shrunken, dry, scaly
Vent	Large, moist, oval; white or pinkish	Small, dry, round; yellow
Pubic bones	Flexible, wide apart; at least three fingers in width	Hard, two fingers or less apart
Abdomen	Soft and pliable, thin-skinned	Firm, thick-skinned
Pigmentation	Pigment bleached from vent, earlobe, beak, and shanks (legs)	Yellow pigment in these areas
Plumage	Bright, smooth, clean	Dull, matted, dry

You will pay the most for pullets, and the least for cockerels, with straight run falling in between.

Try looking for a hatchery in your telephone book, or you can surely locate one in the back of a farm-oriented publication. Another source of day-old chicks is a feed store. Often in the spring, as promotion for its feed, a feed store will give away 25 cockerel chicks of a meat type with the purchase of a bag of feed.

Chicks should be ordered only from flocks that are certified to be pullorum-free. Pullorum-infected chicks (see *Health* section) are often unthrifty and mortality can be quite high. Chicks that are shipped are especially vulnerable, so deal with certified hatcheries. Unless the hatchery is local, the chicks will be sent through the mail. Don't shudder at this—for some reason the never-reliable mail service excels when delivering day-old chicks. (Perhaps the answer to poor mail delivery is to enclose a chick with every piece of correspondence.) After hatching, a chick has enough nourishment for 2 days, which is plenty of time for delivery. The hatchery will notify you a day or 2 before your order is shipped, and miraculously the next day your post office will be filled with the sounds of life. Your next step—immediately—is to put them in a brooder.

Brooding is the process from birth to about 6 weeks of age whereby the chicks are kept confined (but not crowded) in gradually decreasing heat, with plenty of food and water and, above all, excellent sanitation. It is the artificial process similar to the natural brooding you see as a mother hen stands in the yard with a clutch of chicks huddled beneath her. The basic requirements in any brooder setup are:

- Adequate space. One-half square foot per chick up to a month of age and one square foot thereafter until taken from the brooder.
- Correct heat. 90–95 degrees Fahrenheit the first week, after which the temperature should be lowered 5 degrees each week until 55 degrees Fahrenheit, or the outdoor temperature is reached, whichever comes first.
- Adequate food and water
- Adequate ventilation without drafts
- Freedom from predators

I have seen much more elaborate recommendations, including those suggesting infrared heat lamps or gas-fired heating elements; but as long as the basic requirements are met, there should be no problem. For small groups of chicks, we have used a cardboard box with a light suspended in it and placed it in our mudroom. For larger numbers, I built a frame in the corner of an unused room and placed an old screen door on top. (The upper half of the door was screened and the lower half wooden.) I placed a 60-watt light bulb in the lower end and allowed the chicks to regulate the heat themselves by moving closer to or farther away from the light. Their food and water were placed in the screened end. The screening on top let in light and also allowed for good ventilation without being drafty. Another brooder for a large number of chicks is the hover type. A hover with an infrared lamp can either be suspended from the ceiling or set up off the floor with blocks so that it allows the chicks free access to the heat. Heat requirements are the same as listed above, and the chicks can be allowed to use it free choice. Each chick should be allowed 7 square inches of hover space.

Some growers choose to start chicks on newspaper rather than litter in order that the chicks learn to distinguish feed from the litter. After a brief indoctrination period, the gradual transition to all litter can be made. Litter with large pieces, such as wood chips or peanut hulls, is preferred, as it helps maintain even floor temperature, absorbs moisture, is far less dusty, and allows larger pieces of manure to fall through.

In warmer weather, it is essential to check the temperature in the brooder to make sure it doesn't get too hot. During a particularly hot and humid day, turn the light off until the temperature drops in the evening.

Too much light and heat, plus overcrowding in a brooder, can result in cannibalism because of the natural tendency of the chicks. They will start pecking at each other, usually around the eyes and the vent; and, when blood is drawn, all the others will turn on that particular chick and peck it to pieces.

To prevent this, have the chicks debeaked before you buy them. This involves cutting a quarter inch off the top beak soon after hatching. This service is almost always available from hatcheries. If you can't buy debeaked chicks, cutting back on light, heat, and overcrowding should stop the problem. Some people recommend using a light bulb no brighter than 15 watts, but we've always used 40 watts or higher.

At four weeks, you should set a roost an inch or so off the floor in the brooder. It can be a length of sapling, a piece of ½-inch pipe, or an old broomstick. At 6 weeks, when the young chicks are completely feathered out and if the weather is settled, the chicks can be taken from the brooder.

It is not good practice to mix young chicks with older birds. "Pecking order" is a phrase not arrived at by chance. Putting young birds with a group of older birds starts an adjustment period when the older birds dominate the activities of the younger ones, especially at the feed trough. During the period of rapid growth, the young birds should be able to eat all the feed they can without fear of harassment by the older hens. The type of ration being fed to growing chicks is formulated differently from that of laying hens, which is another good reason for not allowing birds of different age groups to commingle until the new birds reach laying age. Continually being pecked at is a very real stress that should be avoided.

One system that works with small flocks is to fence off a separate area in the hen house so that the mature hens cannot come in direct contact. This allows the older birds to see but not molest the growing chicks. By the time the younger birds are close to laying age, they will not be intimidated by the older hens and the hens will no longer view the youngsters as strangers to be picked on.

The housing needs of young chicks are modest and, in good weather, their housing need not be elaborate. A simple shelter fashioned out of two pieces of weather-resistant plywood for a roof and with chicken wire sides will serve nicely. Young birds routinely used to be placed on grass range in this type of shelter and were allowed to run on range until they approached laying age. That system is too labor-intensive for today's larger commercial chicken operations, but remains an option for backyard producers.

Regardless of the size of your operation, good sanitation is essential. Your brooding area should be cleaned and disinfected after the chicks are removed. Similarly, the area where the layers are kept should be cleaned when their laying careers are completed and before a new group of layers is moved in.

Hatching Eggs

This can be the cheapest method of starting a flock if you build your own incubator (see Appendix C) and get fertile eggs from a neighbor. Having a broody hen do the work for you will free you from incubation and brooding problems, but you can't make a hen broody at will the way you can plug in an incubator when you want to hatch a few eggs. Details on hatching eggs can be found in the section on Breeding later in this chapter. Among the predominant breeds of laying hens, the broody factor has largely been eliminated by selective breeding, so finding a broody hen often means going to a bantam or one of the exotic breeds that have been bred more for show than for production.

HOUSING

If you follow to a T plans in poultry books or agricultural bulletins, you will end up with a lovely house that could conceivably be used for a guest house if you ever sell your birds. It will cost accordingly. Use such plans for reference only, and with a little ingenuity and resourcefulness, you'll build a good coop for a fraction of the cost. In some parts of the country, keeping backyard flocks has become a very trendy thing to do, and the housing that these birds enjoy is somewhat over the edge, but the tone of the neighborhood must be kept at an appropriate level, at all costs!

The most important requirements for a good coop are:

- Adequate floor space (at least 3 square feet per bird)
- Good ventilation without drafts
- Adequate lighting
- Safety from predators

It is quite possible to make use of what you have around your property. One of the beauties of chickens is that they demand very little. If you live on an old farm, chances are there is an old milk-cooling shed. In two different places we have lived, we have made use of these old sheds. They convert to fine chicken coops in a few hours (at the most) and will comfortably hold up to two dozen birds. Other potential coops include old tool sheds, erstwhile outhouses, or a corner of a garage or barn. If you keep an eye out, you might notice an old shed on someone else's property. Possibly the owner will part with it for free (or a modest fee), and if it's movable, you're in business.

It is desirable to face the coop into the sun—to the south or southeast. If you are putting your chickens in a barn or other less well-lit location and are not allowing them to run free outdoors, you must supply artificial light. Chickens need 14 hours of light per day to lay optimally. Because laying hens give off a large amount of moisture, you should ensure that there is good ventilation. This can be a door that is left open (except at night) or windows that can be opened. Screened vents along the eaves of the roofline are good if they can be closed off temporarily during very cold weather.

You should have 1 square foot of window space for every 10 feet of floor space. Again, if you are making use of a preexisting structure, you will have to make do with what you have. The windows in one milk shed we converted were mostly broken, so rather than purchase new panes, I nailed chicken wire over the windows and we covered them with clear plastic for protection during the winter. The chicken wire also serves to keep out the predators. Depending upon your area, predators—dogs, hawks, foxes, raccoons, coyotes, weasels, and fisher cats—can lay waste to your flock unless you take measures to keep them out of the coop. Most important of all, close your flock in the coop at night. Since most predators are nocturnal, we allow our chickens to range during the day, and when they go to their roosts in the evening, we simply close the door.

However, there are often good reasons for not allowing hens to roam free, and instead to have a small yard attached to the coop where the hens can be outside throughout most of the year. Hens that are allowed to run freely throughout the gardening season will arouse the ire of the resident gardener. When the garden is started in the spring, chickens are attracted to the freshly turned soil and newly planted seeds. During this time they are better off confined to quarters so they won't end up, literally, in hot water. The option here is to fence off the garden with 3-foot plastic poultry fencing, easy to put up and take down and keeps the hens where they are supposed to be.

For bedding, we use sawdust because it is cheap and easy to get. You can use wood shavings, peat moss, straw, peanut hulls, or ground corncobs. Start with 6 to 8 inches and, instead of cleaning the coop out periodically, let the litter build up, stirring it occasionally, taking out only the soggy spots, and adding more litter every month or so. One method of handling litter is to feed scratch on top of the litter. This encourages the hens to stir and aerate the litter.

You can avoid an unnecessary buildup of droppings on the floor by using one of two common systems. Both will keep the litter in good condition for a longer period of time. One system is to place a platform under the roosts where the droppings will fall onto it. The accumulation can be scraped off and placed in the garden or on the compost pile. (Remember that chicken droppings are very high in nitrogen and may stimulate unwanted growth of some plants.) Another system is to construct a boxed-in pit arrangement under the roosts. This allows you to let the manure build up for longer periods of time. The pit arrangement, while labor-saving until it is time to clean out, serves as a source of ammonia fumes which, if excessive, can be very irritating to the respiratory tract of the hens and the person who draws the short straw

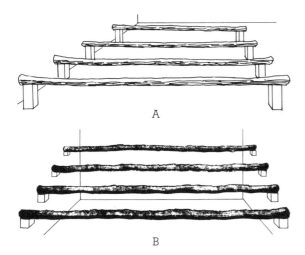

FIG. 1.2: TWO DIFFERENT STYLES OF ROOSTS.
A: Parallel roosts. B: "Stairway" roosts

to clean out the pit. When cleaning a hen house, the individual doing the cleaning should always wear a protective disposable mask to avoid breathing in the potentially irritating dust that goes with the job.

We usually clean out our coop once a year, in the springtime. At that time we have more than 2 feet of litter, and it goes right into the garden because it is one of the best fertilizers we've ever used. After cleaning out the litter, wash the coop with a disinfectant. For some, the idea of cleaning out 2 feet of accumulated litter is not their idea of a good time; if you feel this way more frequent cleanings and a compost pile would seem to be a better idea.

Your coop will need roosts, located away from windows or drafts, and allowing at least 10 inches of space per bird.

You should set them at least 2 feet from the floor of the coop and space them at least a foot apart. They can be placed on the same level with each other (Fig. 1.2a) if you have the room (this allows for better heat retention by the birds in the winter) or diagonally up the wall like treads in a stairway (Fig. 1.2b). Two-by-two lumber is fine, but straight saplings about 1 to 2 inches in diameter are just as good and cost nothing.

To keep the flock under control but allow the chickens to be outdoors, build a run (Fig. 1.3). This allows the chickens limited range as well as opportunities for dust baths and fresh air and sunshine. The run should be attached to the coop so that the chickens are free to leave and enter it. The sides should be 6 to 10 feet high to prevent the birds from "flying the coop." You can get by with 4-foot sides if you're housing heavy meat birds that can't get too far off the ground, or if you clip a number of the primary feathers on the wings of your chickens to short-circuit any dreams of escape. It can be framed with two-by-fours or saplings and covered with chicken wire. Either bury the bottom edge of the wire or place logs around the perimeter of the run to keep predators out.

EQUIPMENT

You will need nest boxes at the rate of one for every four layers (Fig. 1.4). They should be placed in a draft-free area in the back of the coop or wherever they will be least liable to distraction from traffic in and out the door. They should be 14 inches square and a foot deep. They can be filled with wood shavings or straw, and this nest material should be

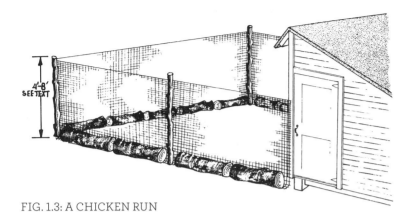

FIG. 1.3: A CHICKEN RUN

changed frequently to prevent soiled eggs. If space is at a premium, they can be stacked as long as they are not out of reach of the hens. I have a cover on the boxes, since the birds seem to like it, and also to discourage roosting on the edges of the boxes with the resultant manure in the nesting material. If, because of the top, a hen has trouble entering the nest, secure a small dowel in front of the box to allow the hen to fly up to it and then make her way into the nest.

You will also need feed and water dishes. They can be purchased fairly inexpensively, but so far I have resisted that urge and my homemade/makeshift feeders have done quite nicely. For feeders, the prerequisites are that the chickens be unable to scratch in them, hence wasting and contaminating feed, and that there is at least 4 inches of space per bird at the feeder.

A simple trough feeder with a reel can be built. If scratching in the feed is a problem, covering the trough with chicken wire so only the beaks can poke through is the answer (Fig. 1.5).

For a waterer, I use an old kitchen pan about 4 inches deep and 8 inches in diameter. I made a little platform to raise it off the floor so that litter is not kicked into it. During cold weather, drinking water must not freeze or there will be a drastic reduction in egg production. I favor hanging a 60- to 70-watt light bulb over the waterer to prevent freezing. This works at even the coldest Vermont temperatures, is much cheaper to buy than an immersion water heater, and draws considerably less electricity. In a small coop, this light can also be used to supply auxiliary light

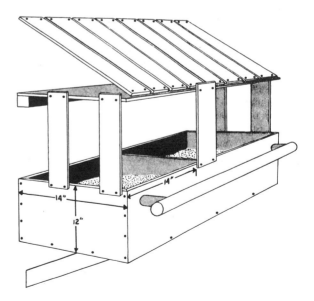

FIG. 1.4: A NEST BOX SUITABLE FOR EIGHT HENS

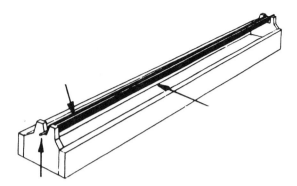

FIG. 1.5: *A trough-type feeder. The length of plastic pipe on the steel rod will spin if stepped on, preventing the chickens from roosting on it and contaminating the feed.*

to maintain the 14-hour days needed for maximum production.

FEED

The diet we have developed for the chicken probably comes closest to that elusive perfect diet that Mother Nature had in mind. With it, healthy chickens for eggs and meat can be raised quite economically on commercial feed alone. However, people wishing greater self-sufficiency can enjoy substantial savings, depending upon the time they wish to spend growing food.

To avoid problems during the chicks' first six weeks of life—the most critical period—it is best to feed a commercial starter mash with 20 percent protein content. At no other age is the furnishing of ample supplies of fresh food and water more important.

In my brooder setup, I feed and water the chicks away from the light. For the first few days of their lives, I spread some newspaper or heavy paper over the litter and sprinkle feed on that, so the chicks learn the difference between their mash and litter.

Thereafter, make clean food available as free choice in chick feeders that you buy or build. (Build the same way you would for larger birds, but proportionately smaller.) I have built troughs on the side of the brooder using pieces of lath, which is even simpler and less expensive. These were 2 inches wide, 2 inches deep, and as long as was needed to accommodate the chicks. The narrowness and closeness to the wall of the troughs made it difficult for the chicks to stand in them and scratch out the feed. If they do, cover the trough with chicken wire.

At first allow 1 inch per chick of feeder space, increase to 2 inches at 3 weeks, and allow 3 to 4 inches per chick at 6 weeks. The greatest saving you can make in feeding is never to fill their troughs more than half full with feed. If you overfill it, they will waste tremendous amounts of feed—and money!

You can buy a chick waterer designed for their use. Chicks by their very nature are inclined to be messy, so save yourself time and aggravation. A properly designed waterer is the way to go rather than using some kitchen discard. Allow a ½-inch of watering space per chick.

Older Birds

You should buy commercial growing mash for chickens 6 weeks to laying age, and laying mash for chickens of laying age and older.

With older birds, as with chicks, have the mash

available at all times, and don't overfill your feeders, or large amounts will be lost in the litter. A pelleted form of mash is available, and chicks waste less of it. If mash is spilled from a feeder, chickens will retrieve very little of it; with pellets, any spilled feed will be visible and cleaned up.

Ranging

I believe in letting chickens "range" as freely as possible during the day. At the very least, you should build a run if ranging is impossible due to predator problems, land restrictions, or the difficulty of finding eggs laid while ranging.

Why exactly is ranging so advantageous? First, the savings on feed is tremendous. I have read that the saving on feed averages 20 percent, but in my experience it has been much greater—50 percent or more. I feed my flock in the morning as I free them from the coop. They may make a few perfunctory pecks at the mash, but are so eager to get out the door and to the earthy

delights that await them that they usually run right past me and my grain scoop. They eat grass, bugs, grit, and whatever turns a chicken on. Another important advantage for us and those of you with other livestock is that chickens clean up after the other animals. With free-ranging chickens, you will have virtually no feed waste. Our chickens clean up after our sloppy horses, grab a few mouthfuls from our obliging sow, and spend a lot of time with their favorite (and best-fed) animals, the feeder pigs. We feed our feeder pigs almost totally on high-quality scraps from a local restaurant. After I empty the feed bucket into their dishes, I clean out the residue in the buckets (maybe a cup or two) with a hose and dump it on the ground. The chickens—which some call dumb, but I wonder—have learned to come running from all corners of the farm at the sound of the hose, and they devour every morsel that is left. In the summer our chickens hardly touch their mash and get along famously on animal "leftovers" and their natural outdoor diet. Free-range chickens fit nicely into one of today's trendiest marketing programs if your operation is large enough to explore the possibility of marketing free-range, organic eggs. If you choose to promote your eggs as free range, use discretion when using the term "organic." It is a regulated term, and those using it must be in compliance with all of its regulations.

Scratch feed

Scratch feed is a nonmash foodstuff that is fed in whole or cracked form by throwing it in the litter or on the ground. The formulas vary widely with some

Table 1.3: Feed Formulas

MARYLAND	LOUISIANA	WASHINGTON	MICHIGAN
Corn 40%	Cracked corn 40%	Wheat 50%	Yellow corn 50%
Wheat 40%	Wheat 30%	Oats 50%	Wheat 50%
Oats 20%	Rice 30%		

common ones listed in Table 1.3. Scratch feed is composed largely of cracked corn with some wheat, sunflower seed, and perhaps some oats mixed in. During the winter months, the high energy level of the corn serves as an excellent source of calories to maintain body temperature during subzero nights. Scratch, when sprinkled directly on the litter, keeps the litter aerated, preventing the litter from becoming packed down and retaining excess moisture.

You can save by growing your own components for one of the above mixtures or you can get by with a scratch feed of whole corn alone or corn with some sunflower seeds added. The grains should be cracked for chicks 6 to 10 weeks old and may be fed whole thereafter. Feed lightly in the morning and more heavily at night. I recommend spreading it in their litter (remove or cover the water bowl and mash feeder, or they will be contaminated as the chickens scratch for feed). It will also help aerate the litter, keep it drier, and expose some of the valuable nutrients in it.

I don't use scratch feed as a regular component of my feeding program. But as Table 1.4 shows, it can comprise up to 50 percent of the feed program, and if you can easily grow it yourself, it is worth considering.

Supplementing Commercial Feed

You can supplement the diet of chickens confined to coops or a run with freshly cut greens (alfalfa, clover, *fresh* lawn clippings, cabbage, kale, Swiss chard, and beet tops). During the winter, fine legume hay, clover, and alfalfa are good sources of feed, as are sprouted grains.

Table 1.4: Suggested Feeding of Scratch Feed

AGE (WEEKS)	4	6	8	10	12	14	16
Mash (percent)	100	95	90	80	70	60	50
Scratch (percent)	0	5	10	20	30	40	50

A simple mix-it-yourself mash, many of whose ingredients you can grow or procure yourself, is shown in Table 1.5. A corn sheller and grain mill, which can be ordered from farm catalogs, are invaluable. If you do it in large enough quantities, it may be economically feasible to have a local grain mill do your grinding.

If so inclined, almost anyone can raise corn and possibly wheat, soybeans, and alfalfa. These constitute the bulk of the feed. Fish meal should be easy enough to make yourself (for that amount you need only be a marginal fisherman); meat and bone meal can come from the offal and bones of other livestock you butcher. If you have a dairy animal, your surplus milk can go into the feed. Even if you have to buy some of the ingredients, you can still save. One note: it is important to grind the feed into a mash, or the chickens will tend to pick out what they like, leave the rest, and not get all the benefits from the mixture.

For the number of chickens I have, I wouldn't think of going to the trouble of mixing my own feed. But for others it may be worthwhile. I depend more on ranging, food from the litter, and having our flock clean up after our other animals. Not surprisingly, this takes less of my time than raising a number of different crops and grinding grain. I do give them any surplus milk we have that doesn't go the pigs, and they get high-quality kitchen scraps (equal to Grade I foods from the section on *Supplementing Commercial Feed* for pigs). I don't feed them in any set proportions, but I experiment, trying to find the best combination of foodstuffs I have available. Watch what scraps you feed, as

Table 1.5: A Simple Mash Mix		
INGREDIENT	POUNDS PER 100-POUND BATCH	POUNDS PER 1-TON BATCH
Yellow corn meal	60.00	1200
Wheat middlings	15.00	300
Soybean meal (dehulled)	8.00	160
Fish meal (65% protein)	3.75	75
Mean and bone meal (47%)	1.00	20
Dried skim milk	3.00	60
Alfalfa leaf meal (20%)	2.50	50
Iodized salt	0.40	8
Ground limestone (38% calcium)	6.35	127
Totals	100.00	2000

Source: *Raising Poultry the Modern Way* by Leonard Mercia. Charlotte, VT: Garden Way Publishing Co., 1975

they are more likely to affect a day-to-day product such as eggs than a more cumulative one, such as pork. In other words, if you fed your chickens a dish of garlic, you would hardly want to have a conversation with one of them in a closed room, much less eat their eggs, I have found a combination of scraps, surplus milk, and ranging cuts our feed bill to practically nothing.

Whenever you are supplementing or substituting for commercial feed, be certain that the flock gets calcium and grit (both found in commercial feed) from other sources. The calcium available from ground oyster shells helps form the eggshells. The grit, picked up naturally by free-ranging birds, enables the toothless chicken to grind its food. For birds not raised free range, a calcium ration and grit should be available if commercial mash is not used for feed.

MANAGEMENT
Routines

The routine a person will want to develop in managing his or her flock will take time, and some mistakes will be made before it suits the family's needs.

A hen will begin laying walnut-sized eggs anytime from 4½ to 6 months of age. Soon the eggs she lays will be normal size. As my young chickens reach maturity (also signaled by some rather discordant crowing from the young roosters), I usually butcher off all but two of my roosters, keeping a spare in case of tragedy, keeping the *best* for breeding, and resisting the temptation of imagining how much better they would look on the dinner table.

For a hen's first year, you can expect an *average* of an egg a day during her productive period (see

Molting section). Depending upon the breed, she will lay from 200 to 240 eggs during her first year and decreasing numbers—though larger eggs—in succeeding years.

You will have to decide whether to keep your hens for more than one year. After the first year, when production stops, the question is whether to keep these at a less productive feed-to-egg-conversion ratio or spend the money to buy and raise new chicks to laying age. This will depend largely on the price of new chicks in your area and the amount of feed you need to purchase, or whether you can supply it through your own devices.

If you wish to keep your hens another year you will have to accept the molt with understanding . . . and no eggs. I don't enjoy buying eggs, so I have worked out a routine that enables me to get around molting economically. We started our flock with pullets and after a summer of production, a molt, and a practically eggless winter, we hatched out 25 chicks the following spring. By the time the original birds went into molt again the next fall, the chicks had begun laying. We butchered the older birds and took the new layers through the winter (allowing 14 hours of light and plenty of thawed water so as to not upset production) without a molt, and hatched another batch of chicks that spring. Again, we butchered the older hens that fall as the new chicks began laying. In this way we never had our hens go through molt, had plenty of meat, and never went through gaps in egg production.

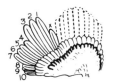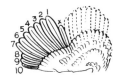

FIG. 1.6: *Wings during different stages of molt. (1) shows the 10 old primary feathers (black) and the secondary feathers (broken outline), separated by the axial feather (x). (2) shows a slow molter at 6 weeks of molt, with one fully grown primary and feathers 2, 3, and 4 developing at 2-week intervals. In contrast, (3), a fast molter, has all new feathers. Feathers 1 to 3 were dropped first (now fully developed); feathers 4 to 7 were dropped next (now 4 weeks old); and feathers 8 to 10 were dropped last (now 2 weeks old). Two weeks later (4), feathers 1 to 7 are fully grown. Fast molt took 10 weeks, compared to 24 weeks for slow molt. See text for further explanation. (South Dakota Extension Service drawing.)*

Molting

At the end of a laying year (or at times due to disease or mismanagement), hens molt, or drop their feathers, in order to renew them (Fig. 1.6). They look sickly, don't lay (some do, but these are rare), and to top it all off, tend to eat more. Depending on the bird, the molt cycle (dropping and growing feathers) can take from 10 to 24 weeks, with little or no egg production during that time. Those taking longer to molt should be culled from the flock. Fast molters often lay for a few weeks after dropping feathers, so egg loss is at a minimum. They tend to drop feathers in clumps rather than singly, and grow them back in clumps, hence returning to production more quickly. Slow molters drop one feather at a time and renew them one at a time. They often begin molting earlier and continue for months. Eat those birds.

When molting begins, feathers are shed in the following order: head, neck, breast, body, wings, and tail. Some birds will molt after only 8 or 9 months of production (early molters—cull them), while others will lay for 12 months before the onset of molt. The primary feathers of the wing (see Fig. 1.6) are used to chart the molt, because they are dropped and renewed in a set order. By examining the primary feathers, you can estimate how long a bird has been in molt and how soon before she will return to laying. If the feathers are dropped singly or in groups, you can decide if the bird is a slow or fast molter. The first primary feather to be dropped is in the middle of the wing (Number 1 in the upper left corner of Fig. 1.6) and it takes 6 weeks for a new primary to be fully regrown. Normally, the next primary drops in 2 weeks; the next, 2 weeks after that, and so on. When a fast molter drops 2 or more primaries at a time, they will also grow out together. By dropping primaries in groups of two or more, these fast molters (and high producers) complete a molt more rapidly.

Force Molting

For birds that do not normally go into molt after 12 months, or for those birds whose production lags prior to molting, it may be desirable to use force molting (Table 1.6). Through this procedure, the birds will drop their feathers and will return to production after new feathers are grown, which may be as soon as eight weeks. Egg production, as it normally would be in the second year, is lower after force molting. In addition, the eggs are larger, feed consumption is higher, and the health of the flock will be poorer. I have never used it, because unlike a commercial poultryman, I do not live or die with relatively minute changes in production, and I also consider it an unnatural strain on the birds. It is, however, an alternative to buying a new set of birds after the first year of production.

Culling

Culling is the removal of unhealthy or poor producers from a flock. Unless diseased, a cull bird offers the compensation that it is at least as good for the table (maybe meatier due to lack of strain from laying) as a layer. Eat well.

Egg Production

Maximum egg production cannot be maintained without thawed water in the winter and 14 hours of light per

Table 1.6: Force Molting Procedure

(A) FOR NORMAL WEATHER CONDITIONS

Day 1	Decrease light to eight hours per day in light-tight houses, or no natural light in window houses. Remove all feed and water.
Day 3	Provide water.
Day 8	Provide growing mash—40 percent normal consumption level.
Day 22	Restore lights to premolt level and offer full-feed laying ration.

(B) FOR HOT WEATHER

Day 1	Decrease light as above. Remove feed but allow water free choice.
Day 8	Provide growing mash—40 percent normal consumption level.
Day 22	Restore lights to premolt level and offer full-feed laying ration.

Source: *Raising Poultry the Modern Way* by Leonard Mercia. Charlotte, VT: Garden Way Publishing Co., 1975.

day (Table 1.7). As the days grow shorter, artificial light must be supplied so that the 14-hour requirement is met. Lights can be turned on before dawn or after dusk, or in combination. Since I will not buy an electric timer to control my lights (it costs too many dozen eggs), I have found that putting the light on at dusk and turning it off before bedtime is most convenient for me.

Allowing laying chickens to range freely over the premises may be delightful to them and help you to save on feed, but it can wreak havoc with gathering eggs. If not confined, chickens have a habit of finding secluded spots to lay their daily eggs. Often the owner is unaware of these sheltered areas. It is a frustrating experience to find a large batch of eggs in a corner of the barn and try to determine their ages. Man has yet to develop a chicken that lays a date-coded egg. Pick up eggs at least once a day, and twice a day in cold or hot weather. This serves another function: it helps to avoid the problem of hens eating their own eggs.

Sometimes this urge to lay eggs in unexpected places is due to a filthy nest box, but more often it is the capriciousness of the birds. If I can find out where they're laying (they love haystacks and similar out-of-the-way places), and I can collect them easily, I usually don't bother the hens. If not, I try confining them to the coop for a few days so they have no choice.

Caponizing

Caponizing, or castrating the male chicken either surgically or chemically, is beyond the scope of this book. Leonard Mercia's book, *Storey's Guide to Raising Poultry,* goes into considerable detail about this subject. If you want to raise them—and they make good eating—capons can be bought from most local poultry farms. They cost about four to five times as much as day-old chicks, and they are sold at about 4 to 6 weeks.

Handling

The best way to carry a chicken is to hold it upside down by the feet. The blood rushing to its head will subdue it after a few moments of indignation. Unless you lock up your chickens at night and can corner them in the coop, catching a chicken is another story. I usually wait for them to put themselves to bed at

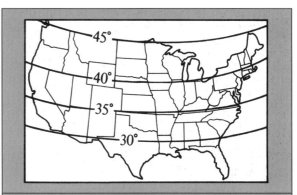

Table 1.7: Hours of Daylight by Month

MONTH	45° N. LAT.	40° N. LAT.	35° N. LAT.	30° N. LAT.
January	9:09	9:39	10:04	10:25
February	10:26	10:41	10:56	11:09
March	11:53	11:54	11:57	11:58
April	13:29	13:15	13:04	12:53
May	14:51	14:23	13:59	13:39
June	15:35	15:00	14:30	14:04
July	15:07	14:34	14:07	13:44
August	14:06	13:46	13:29	13:14
September	12:33	12:30	12:26	12:22
October	11:01	11:12	11:20	11:28
November	9:34	9:59	10:21	10:41
December	8:48	9:22	9:50	10:13

You must supplement daylight to maintain 14 hours of light per day to obtain maximum egg production. Refer to the map and the chart to determine how many hours of supplementary light (if any) you must supply. More accurate readings may be gained through interpolation.

A totally enclosed poultry yard offers good protection against both flying and climbing predators

night, or grab one before letting them out in the morning. I have inadvertently caught one in a Havahart trap when I was trying to catch rats. Poultry hooks that snag a leg, available from most farm catalogs, make what is potentially a day-long task a snap.

Predators

Many predators can threaten your flock, both during the daytime and at night.

Rats and mice are often the greatest nuisance. They love chicken feed, and will eat it as well as destroy the bags it is kept in. There are many methods of getting rid of them. Keep the feed in ratproof containers. Eliminate the hiding places they prefer, such as the spaces between double-sheathed walls or under floors, or in stored equipment. Set traps for them. Keep a cat. If all else fails, put out anticoagulant bait. To avoid environmental damage or accidental poisoning of other animals, use a proper bait box that only rats and mice can enter, or bait blocks that can be deposited deep into holes.

Dogs will raid a henhouse or a flock outside, killing some and causing many of the rest to pile up and suffocate. It's difficult to break a dog of this (to them) sport, particularly if it's a neighbor's dog. Try a person-to-person appeal to have the dog tied up before you explore legal options.

Cats are troublesome only with small chicks. Keep the chicks well protected. Hawks will strike in the daytime, carrying off young birds or eating larger ones on the ground. Owls, too, will use much the same approach. A yard enclosed with a high fence offers some protection.

Many wild animals will dine on chicken given the opportunity, even eating inside a henhouse to do it if you offer an opening. These include foxes, skunks, minks, weasels, and raccoons.

BREEDING

While you can get eggs with just a flock of hens, to get fertile eggs you need a rooster. Hatching your own eggs enables you to raise replacements and cut your ties to hatcheries. One rooster can handle up to 15 hens (note: 14 hours of light a day are also essential for maximum fertility of your rooster). As a rooster is routinely pecking about the yard he will, as if a button on a distant control board was pushed, suddenly look up, begin chasing a hen, collar her by the neck, drive her face into the ground, and mount her. It is hardly the tenderest of nature's mating rituals.

A permanent chicken house built on pilings above the ground to discourage rats and other vermin from nesting and climbing to steal eggs or birds. It is ideal for the small backyard flock.

Artificial Incubation

Incubators can be purchased from feed stores and ordered from farm catalogs, but unless you plan to hatch eggs on a large scale, I doubt the economic value of such an operation (and that's presumably why you'd choose to hatch eggs in the first place). It is possible for anyone to construct an incubator with a minimum of time and money (see Appendix C).

If you are collecting fertile eggs day by day, you must keep them under carefully controlled conditions until you incubate them. Storage of up to a week has a minimal effect on an egg's hatchability, but after 12 days the viability declines quickly. Eggs must be kept at 55–65 degrees Fahrenheit at a humidity of 70–80 percent. Under these conditions, the egg cell remains dormant. At 80–90 degrees Fahrenheit, the cell germinates but soon dies; at temperatures above 110 degrees Fahrenheit, the cell will also die. Eggs should be stored small end down at an angle of 35° and rotated twice each day. Discard any eggs that are misshapen, cracked, or unusually large or small.

An incubator must be set at a temperature within the very strict limits of 99–105 degrees Fahrenheit. Start the incubator the day before to check its operation and to allow it to heat up before inserting the eggs. Warm the eggs up to room temperature before placing them in the incubator so their temperature will not rise sharply when they are put in. Have a small pan of water on the floor of the incubator to prevent the air from becoming too dry. Finally, turn the eggs three or four times a day. Make a small mark on one side of each egg to avoid confusion when turning.

If you have done everything correctly, in 21 days (give a day or 2 either way) you should notice cracks appearing on the eggs and perhaps hear little peeps coming from them. The breaking-out process is a long and arduous one for the chick and can take up to a day. If after that time a chick hasn't emerged from the egg, it is generally recommended that you leave it and let nature take its course.

However, I can never bear to see a beak poking out and leave the chick to die, so I usually help the little fellow out. Thus far, I have never lost any chicks this way, and my henhouse has not been struck by lightning from above.

In planning the ultimate size of your flock, figure on an average hatch of 60 percent. I usually raise my chickens for both eggs and meat, so I am not unduly concerned with the sex of the chicks. I just butcher off most of the cockerels when they reach the size I choose to use them.

Once a chick hatches, keep it in the incubator until it is dry and fluffy, about 12 hours. From then on, manage as you would a day-old chick that you've purchased.

Natural Incubation

I find this method much easier because the mother hen will do the brooding and you don't have to build an incubator. On the other hand, a hen will not hatch eggs whenever you want her to. This method is less efficient, too, because it puts a bird out of production for the hatching and brooding period.

All breeds are not equally suited for mothering. Some breeds have had the brood characteristic bred out of them. Bantams, Wyandottes, Plymouth Rocks, and Rhode Island Reds make good mothers and will generally set, especially as they get older. If you want chicks, keep an eye out for a broody hen and place some eggs under her if she doesn't have enough of her own. A large hen can cover up to 15 eggs, and a medium one 9 or 10. Make sure she isn't disturbed and have water and feed nearby for her infrequent feedings.

I leave my setting hen in the coop, rather than moving her, as unfamiliar surroundings may break a hen out of her broodiness. If you notice other hens crowding her out of the box and laying too many eggs with her, you probably don't have enough nest boxes, or the others are soiled. Once she sets, she will take over from there. She turns the eggs every day, and when they hatch, she will brood them until the chicks are old enough to go out on their own. Often there is a problem of acceptance of the mother and her clutch by the other birds, but normally after a few weeks the new family will become accepted. At times we've had chickens become wild, roosting in trees and returning to the coop only for food. In this case, lock them in the coop for a week, and these chickens should become as domesticated as the rest of the flock.

HEALTH

In combating diseases, your most effective tactic is prevention. Letting your chickens range free to reap the benefits of fresh air and sunshine is one way to promote good health. Disease in my flock, like most small flocks, is not a major problem. Disease is more likely to occur in confined commercial flocks, which are managed in a much more intensive way. Signs of ailing birds are a drop in feed and water consumption, lower egg production, and listlessness. Unless you are going into large-scale production, the purchase of a veterinary text devoted to chickens is unnecessary. The disease table in the Appendix will help.

An excellent source of information relating to poultry diseases is the department of veterinary science at your state college of agriculture. When you need advice or treatment recommendations, a call to this source will result in a wealth of information.

Table 1.8: Butchering Terminology for Chickens		
TERM	**SEX**	**WEIGHT OR AGE**
Broiler	Either	2½ lbs. (Less than 8 months old)
Fryer	Either	2½–3½ lbs. (Less than 8 months old)
Roaster	Either	3½–5 lbs. (Less than 8 months old)
Capon	Castrated male	6–8 lbs.
Fowl	Female	Over 8 months
Cock or stag	Male	Over 8 months

If you have unexpected losses, presenting a dead bird will often provide the information the poultry pathologists need to answer your questions.

BUTCHERING

Chickens are excellent animals for starting your livestock butchering operation, as they are relatively easy to butcher (Table 1.8). When you decide to butcher a bird, you should deprive it of food, but not water, for 12 hours.

When catching a hen for butchering, watch your roosters. They are never pleased to lose a member of their harem and can get quite nasty. Our rooster clucks indignantly as I stalk one of his lovelies and

FIG. 1.7: NEVER-MISS CHOPPING BLOCK

then attacks me after the catch, but only after I turn my back. I can usually outrun him.

Whatever your choice of butchering size and age (and whatever the personality of your rooster), you will have two methods of killing. A third, breaking the neck, is rarely used. The most popular are beheading and sticking and cutting the throat. Both are equally effective.

Perhaps beheading is best for the beginner, since it is the quickest and surest. Holding the bird by the legs, I tie some bailing twine around its feet so that it can hang for bleeding after I've cut its head off. (By hanging it this way you will also be spared the anguish of your first chicken flapping madly about the yard without a head.) Have a sharp ax and a chopping block (a stump will do) with two nails driven in and sticking up 2 inches parallel to each other and about 1 inch apart (Fig. 1.7). Insert the chicken's head between the two nails and pull gently. The beak will get caught between the nails and the neck will stretch, giving you a surer shot. After chopping the head off, hang the bird up immediately and let it bleed out completely. I usually let it hang

10 minutes or so; much longer, especially in cold weather, will make the bird harder to pluck.

When sticking, tie the bird's feet and then hang it upside down. Using a sharp, short-bladed knife, cut the throat at the base of the neck, being sure you sever the jugular vein. Immediately insert the point of the knife into the bird's mouth and force it through the roof toward the back of the head into the brain cavity, and give the knife a quarter turn. The piercing of the brain causes a squawk and a convulsive wing flapping that loosens the feathers and makes the plucking easier.

Plucking

Plucking may be done by the wet or the dry method. Begin dry plucking immediately after the bird is bled and before it cools, since this tends to set the feathers. This is best done after sticking, rather than beheading. For wet plucking, dip the bird in hot (150–190 degrees Fahrenheit) water for 30 seconds to a minute. Too long and the skin will cook. If you wish to freeze your bird for any length of time, it is best to dry-pluck since scalding tends to cut down on freezer life.

Plucking is hardly fun, and after doing a few birds, you may be ready to switch to rabbits for your source of meat, and raise chickens just for eggs. Your skill and speed at plucking will improve with practice. It is best to take your time at first so you will not tear the skin and thus expose the meat to drying. The best order of plucking is to do the wings, breast, body, back, legs, and finally the neck. If any small pinfeathers are left, they may be removed with a dull knife. Small feathers and hair may be singed off with a gas flame or a candle.

Dressing

It is best to work with a chilled carcass. A sharp knife is essential. Poultry shears are helpful but not necessary. First cut off the head and, peeling back the skin, sever the neck and save it for the stockpot. Then remove the feet at the hock joints. (After washing, the feet make an excellent addition to the stockpot.) Make a cut in the vent large enough to accommodate the hand. If the bird is of laying age you may find a fully formed egg and see a string of yolks in varying stages of development. Pull out the innards, saving the heart, liver, and gizzard. Carefully remove the gallbladder from the liver and remove the inner sac from the gizzard. Push your fingers in the front and remove the lungs, or "lights." Flush with water and make sure all pieces of blood clots are removed. After the bird is washed thoroughly, it may be roasted, cut up for frying or broiling, or frozen.

Ducks

Ducks are hardier than chickens. They require less attention and they grow more quickly. While they do not forage for food as much as geese, they will still augment their diet by ranging, and they will produce tasty meat very economically in only eight to ten weeks.

Ducks need very little space and almost no housing. Your ducks will not require a pond or a brook, but if you have one, they will be happy birds.

The meat is delicious, not at all gamy, though with more fat than wild duck. You also have the opportunity to gather duck eggs and even some duck down for pillows or whatever else you may choose to do with it.

Ducks have something else over chickens. They have personalities! Few people get pleasure from watching chickens, as they have the maddening tendency to behave the same way all the time. Ducks frolicking on a pond or waddling mechanically in a line to some distant object are a sight to behold. Be careful: you can make ducks into pets (try that with chickens), but if their ultimate destination is your dinner table, this can put a crimp in your plans.

BREEDS

Some breeds of ducks are well suited to meat production and others are egg producers. Unlike chickens, there is no such thing as a dual-purpose breed of duck. You can eat good egg layers, but the car-

FIG. 1.8: EXAMPLES OF DUCK BREEDS

casses will be small and of poorer quality, since most of their energy goes into the production of eggs, not meat. The meat birds will supply you with a trickle of eggs along with meat. Duck eggs, while larger than chicken eggs, are good for baking, but many people find them unsuitable for "straight" eating. If unfamiliar with their taste, try them before you decide to live on them.

Meat Breeds

White Pekin: This is the Long Island duck you've heard about and is the major breed raised commercially in the United States (Fig. 1.8). Pekins are large birds with white plumage, orange-yellow bills, reddish-yellow legs and feet, and yellow skin. Adult drakes (the males) reach a weight of about 9 pounds, while adult ducks (the females) reach about

8 pounds. They produce excellent meat and reach market weight of 7 pounds in 8 to 10 weeks. They are also reasonably good egg producers, with a duck laying an average of 160 eggs per year. They are high-strung birds and, as a result, are poor setters. If you wish to hatch their eggs, it must be done by artificial incubation or by enlisting the services of a broody hen.

Muscovy: While there are many varieties of this native South American, the white ones, because they dress out more attractively, are most desirable for meat production (Fig. 1.8). They have white skin, and the darkening around the eyes resembles a mask. Adult drakes weigh 10 pounds, and ducks weigh about 7 pounds. They reach market weight in about 10 to 17 weeks. They are poor egg producers (average 40 to 45 eggs per year) but they are the best setters of all meat producers.

Aylesbury: This is the English counterpart of our White Pekin, and its quack has a characteristic British air. Like the Pekin, they reach meat weight (7 pounds) in 8 to 10 weeks. They have white feathers, white skin, flesh-colored bills, and light orange legs and feet. Adult drakes weigh 9 pounds, ducks 8 pounds. While they are not as nervous as Pekins, they do not have much interest in setting. They are not quite so prolific in egg production as Pekins.

Other meat breeds include Rouen, Cayuga, Call, and Swedish.

Egg Breeds

Khaki Campbell: Khakis are able to surpass even the highest-producing chickens in egg production. They are, however, not valued for their meat, since drakes and ducks reach a mature weight of only 4½ pounds.

The Khaki drakes are bronze on the lower back, tail coverts, head, and neck; the rest of the bird is khaki-colored. Their bills are green and they have orange legs and feet. The females have seal-brown heads and necks; the rest is khaki-colored. Their bills are greenish-black and their legs and feet are brown.

Indian Runner: Indian Runners fall short of Khakis in egg production, but they are good producers (Fig. 1.8). There are three types: White Penciled, Fawn, and White. All three have orange to reddish-orange feet and legs. They stand quite erect, their carriage being almost perpendicular to the ground. Their adult weight is about that of Khakis. They are not valued for their meat.

PURCHASE

As with chickens, you have three choices: hatching eggs, buying day-old ducklings, or buying older stock for breeding. Your choice will depend on how you will raise your birds. If you plan to have only a few ducks for meat, buy a few day-old ducklings from a local hatchery or friend and raise them. This frees you from getting incubating equipment and fussing with breeding and hatching. If a larger flock is desired, you can buy fertile duck eggs or buy a lot of ducklings and go from there, selecting the best for future breeding stock.

Hatching

The hatching time for duck eggs is 28 days for all breeds except Muscovy, which is 35 days. You can use the same incubating and brooding methods as for hatching chickens. The eggs should be placed in the incubator small end down, or if that is impossible, rest them on their sides.

For natural hatching, Muscovys are the best breed to use, and frequently the only ducks that will hatch eggs. If you don't have a setting duck, you can use a broody hen. A hen will handle up to ten duck eggs (don't let her take more than she can handle, since this will affect the hatchability of all of them); a duck, slightly more. Sprinkle warm water on the eggs each day if you are using a hen or a duck that has no access to water, since duck eggs need more moisture than chicken eggs do. When they hatch, in the case of a duck mother, let nature take its course; in the case of a hen foster mother, confine her and her brood of ugly chicklings so that she cannot wander off as they will not be able to keep up with her the way chicks do, and they may get lost. After four weeks, the ducklings should be able to keep up with her and she can be set free.

Day-Old Ducklings

In purchasing day-old ducklings locally, you have the advantage of being able to buy fewer (mail-order minimums are often 25), but you may not be

able to have your choice of breed. In any event, you will need a brooder. Handle brooding as with chicks, but allow 1 square foot of space per bird. Ducklings normally need brooding for only 4 weeks and may even get by with 2 or 3 weeks. In warmer summer weather, a light bulb in their shelter may suffice. Do not allow a duckling to get soaked by rain before 4 weeks of age or until it is feathered. It should also not be allowed to swim before 6 weeks, because the oil is not fully distributed on its feathers, and it will be more susceptible to chills and sickness. Shade must also be available to young birds.

Older Stock

This applies to choice of mature stock which are bought for breeding or to picking potential breeders from your own flock. At 6 to 7 weeks, drakes and ducks can be distinguished not only by the difference in coloration (males have brighter plumage), but also by their sounds. Females "honk" and drakes "belch." You will require one drake for every six females for

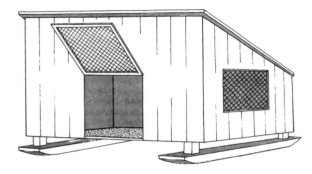

FIG. 1.9: A MOVABLE HOUSE FOR DUCKS OR GEESE

breeding. Look for ducks with evenly colored feet, legs, and bills. Choose birds that are vigorous, alert, heavy, and solid, with broad breasts, and necks that are not too long. Plumage should be even and glossy. Find out, if you can, the breeding records of the parents; choose ducks from parents that have exhibited high fertility, hatchability, and egg production.

Predators

Predators are generally the same as for chickens. Your dog, while perhaps accustomed to chickens, may see the newcomers as fair game. Watch him.

HOUSING

If you plan to raise ducks for meat in the summer months, housing is quite simple. If you don't have a handy pond or stream, you can house them with the chickens—if they will stand for it. If the chickens peck and chase the ducks, as chickens may do, make a little pen within the coop that only the duck-

lings can scoot under. Within a few weeks they will probably be accepted, but if not, they will be close to butchering age anyway and out of your and your chickens' hair (or feathers).

Allow ducks to range free all day, then lure them with grain into a small enclosure and lock them in at night to safeguard them from predators and hungry neighbors. A simple, movable hutch should have a dry floor and furnish protection from severe weather and predators (Fig. 1.9). They don't need roosts, but allow 5–6 square feet of floor space per bird.

If you wish to raise ducks, you will need a more permanent house. It can be much the same design as a chicken coop, but no roosts are needed. Ducks are more resistant to cold than chickens, so you don't need to be so careful about plugging up drafts in the building. However, you shouldn't allow any snow or rain to seep in, and the house should, of course, be

Allow ducks to range free during the day.

predator-proof. Space requirements are the same as for the temporary shelter described above. If feather pulling is a problem, the ducks may be overcrowded. Give them more space. If this doesn't work, debill them (clip off the forward edge of the upper bill). Keep the house scrupulously clean. A built-up system works well. Be sure to shovel out any wet spots.

If you allow your birds to range, as I strongly suggest, they will be healthy and disease-resistant. Keep them away from pools of stagnant water, since these are havens for disease. If for any number of reasons (limited space, predators, large flocks) you must fence your birds in order to let them run, allow at least 40 square feet per bird. Keep the run very clean. Because ducks can be quite messy, sand is the best ground cover, and it helps if you locate the run so that it slants downhill away from the house. For run-confined birds, you must throw in fresh greens daily to realize any feed economy.

EQUIPMENT

Ducks do not need swimming water, nor is it necessary for fertilized eggs, but if you have a pond, your ducks will love you for it. Don't allow the population to get too high or your pond will be a mess and will smell awful. For ducks confined to a run or on a pondless farm, supply them with a pool they can splash in and clean themselves. This is especially good if you have ducks hatching eggs, since it will enable the duck to supply more moisture to the eggs. The pool should measure at least 3 feet by 1 foot and

be deep enough so the ducks can totally immerse themselves. It can be constructed out of concrete or wood, or you can use an old feed trough or discarded bathtub. Change the water frequently to prevent stagnation.

Feed Dishes

Your ducklings can be fed from chick feeders or shallow troughs or pans. Since they do not scratch their food as chickens do, waste is less of a problem. As they get older, you can feed them from larger troughs or from a self-feeding hopper (Fig. 1.10). This hopper is very useful if your birds are free-ranging since you don't need a larger house to hold the feeding equipment. If you have free-ranging chickens, food piracy may be a problem. We have handled this by separating them from each other's ranging area.

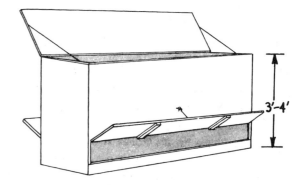

FIG. 1.10: A SELF-FEEDING RANGE FEEDER FOR DUCKS OR GEESE. *Lower panels may be closed at night to prevent piracy by other poultry or rodents.*

Waterers

The duck's waterer must be constructed so that the whole bill (but not the whole body) can be immersed and the nostrils can be cleaned. Ducks are very messy drinkers, so if they are watered inside their house, place the waterer on wire off the floor and have a lot of absorbent litter underneath. Better yet, don't give them their water while they're cooped up. They can adjust to being without it while closed in at night and their egg production and growth will not be affected. If you keep the waterer outside, the food must be kept out of the coop, too, since a duck eating dry food without access to water may strangle.

Nest Boxes

If you keep ducks for egg production or breeding, nest boxes must be provided. They can be constructed like chickens' nest boxes, but should not have tops. They should be a bit larger, at least 15 inches square and 18 inches deep, and they should be set on the floor. Nest material can be the same as for chickens.

Stampede Lights

When startled at night, ducks have a tendency to run in circles, possibly causing injury or death. A low-wattage nightlight will help keep this from happening.

FEED

Ducks benefit (or maybe it should be said that you benefit) from an excellent feed conversion ratio (3:1),

Table 1.9: Feeding Rations for Ducks
Starter for the first 2 weeks (20–22 percent protein)
Grower for meat birds, 2 weeks to market (15–18 percent protein)
Breeder developer for breeding and egg flock, 6 to 7 weeks of age until 1 month before production
Breeder from 1 month prior to production

and this should enable you to raise meat and eggs economically. While they are not as good at foraging for food as geese, ducks will augment their diet if they are allowed free range, or supplied with greens if penned.

Table 1.9 outlines the rations that should be fed to ducks:

Pelleted feed, if available, is preferred to mashes, as there is less waste and the ducks convert it to meat at a better rate. Most feed stores should carry duck feed or feed that is suitable for them. Chicken feed can be substituted for growing flocks, but *medications present in chick starter can kill young ducklings,* which should be fed either an unmedicated chick starter or feed formulated for ducklings. The starter feed can be in the form of pellets (no larger than 1/8 inch) or mash that is slightly wet. Allow the feed to be consumed free choice and supply fresh water.

For meat ducks, allow free choice all the grower feed they will eat. Again, mash can be used, but a 3/16-inch pellet will be utilized with greater efficiency and less waste. The ducks you plan to use for breeding should be fed the breeder developer ration. To prevent them from becoming too fat, feed this on a limited basis, ½ pound per duck per day, split between morning and evening feedings.

At one month prior to egg production (explained in section on Breeding), switch the ducks from the breeder developer ration to a breeder ration and feed them free choice. If feed does not contain a calcium supplement (which is unlikely if it's a commercial feed), ground eggshells—chicken or duck—should be added.

Supplementing Commercial Feed

While their feed conversion ability is good, and their relatively small level of feed consumption makes supplemental feed less important than for pigs, you can still make some savings. It is impractical to mix your own feed, but you can augment their diet by allowing them to free-range. They will obtain a great deal of their feed requirements naturally and balance their diet at the same time. Ducks that are penned should be supplied with fresh greens at least once daily.

BREEDING

Ducks that have hatched from April through July will reach maturity at 7 months. Those hatched between September and January will mature in 5 or 6 months because of the increasing length of the days in their growing period. It is best to pick your future breeding stock when they are 6 or 7 weeks of age and begin feeding them as indicated in the Feed section. (The sexing of ducks is discussed in the Purchase section and under Breeding in Part III, Geese, which follows.) You will need one drake for every six ducks, but I suggest at least one extra in the event that one of the drakes meets an untimely end.

When your birds are near maturity, and 3 weeks before you wish them to begin full egg production, increase the light in their coop so that they receive 14 hours of light a day. Give the drakes an extra 1 or 2 weeks' head start so you'll be sure they're ready for the ducks. Bringing females into full production (supplying them with 14 hours of light) is not recommended before seven months of age because they will tend to produce smaller eggs with a lower hatchability.

Once the duck begins laying, she will reach her production peak within 5 to 6 weeks. If you are interested in hatching eggs, fertility is highest when egg production is highest, and the greatest hatchability will occur after the first few settings, probably because the birds will have improved mating proficiency. The details of handling fertile eggs, incubation, and brooding are given in earlier sections.

Ducks lay most of their eggs before 7 AM, so keeping them locked up until after they are through laying will not markedly affect their feed consumption (if you keep the feed outside) or their ability to range. The number of eggs they lay will, of course, depend upon the breed. High production will be maintained for 5 to 6 months and then will taper off gradually, allowing the ducks a rest period.

HEALTH

Prevention is the byword. Clean pens and runs frequently and avoid crowding, and you shouldn't even need to consult the disease chart in the Appendix. Keep young ducks from becoming chilled and keep all ducks away from stagnant water.

BUTCHERING

Ducks reach market weight at 7 to 10 weeks of age (except Muscovy: 10 to 17 weeks). Butcher those with firm, plump carcasses without many pinfeathers. Raising them beyond the recommended ages and weights will perhaps produce larger ducks, but not so economically.

Keep ducks from food for 12 hours before slaughter, but allow access to water. Methods of killing are the same as for chickens. You can dry-pluck or scald; wax on the pinfeathers may be quite helpful.

Geese

Geese are the hardiest of all poultry; they are the most disease-resistant and the cheapest to raise. Because they are such good foragers and can be set out to forage when they are young, they can be raised with an absolute minimum of feed. They also make good "watchdogs," a plus in crime-ridden suburbs. Walk a goose every night, and see if you get mugged. They are good weeders, their feathers and down can be sold for extra cash, their meat is excellent (and abundant).

And, yes, they are very entertaining. To me, they are the reincarnation of an especially "rammy" ram. People we know have geese that were especially playful and, in fact, enjoyed playing with their children. One day as friends were visiting and they were riding their snowmobile, one of the geese jumped on the back for a ride. Naturally, the friend, upon seeing this, jumped off. Away the goose rode, honking madly, across the field and into a ditch. They may also attack strangers if they come into "their" territory. To the unsuspecting visitor, the sight of a hissing goose coming at them with its wings fully extended can be a rather intimidating experience. They are not above giving their target a good peck on the leg, and that is not a gentle tap.

BREEDS

There are breeds of geese to suit your various tastes: meat, eggs, eggs and meat, show, and pets. Table 1.10 gives the major breeds of geese and their sizes, as

Table 1.10: Breeds of Geese and Their Weights in Pounds				
	MALE		FEMALE	
BREED	YOUNG	ADULT	YOUNG	ADULT
African	16	20	14	18
Buff	16	18	14	16
Canada	10	12	8	10
Chinese	10	12	8	10
Egyptian	5	5½	4	4½
Emden	20	26	16	20
Pilgrim	12	14	10	13
Sebastopol	12	14	10	12
Toulouse	20	26	16	20

recommended by the American Poultry Association in the *Standard of Perfection*.

African: The African has an erect body that is carried high from the ground. Its head is light brown with a large black knob atop. It has large dark brown eyes and a black bill. The feathers are light ash brown on the underside of the body and on the neck and breast, and a darker ash brown on the wings and neck. African geese are good layers and they grow rapidly to market weight, but they have more pinfeathers than most meat birds. They also tend to be very noisy.

Buff: This breed has poor meat and egg-laying qualities and is not worth considering for our purposes.

Canada (wild): These geese are not good meat or egg birds and thus are valuable only for pets or show. You need a permit to own one. Canada geese can be kept confined only in a wire pen or by clipping their wings. They are better off left in the wild.

Chinese (brown or white): This is a good multi-purpose breed. Its graceful, swanlike appearance makes it popular for show or pets; its relatively high egg production (40 to 65 eggs per year) makes it a desirable egg bird (Fig. 1.11). It is of medium weight, grows rapidly, matures early, and its meat is delicious. As with all poultry, the white variety is more desirable for meat purposes because of the absence of dark pinfeathers. Chinese geese also make very good "watch dogs."

Egyptian: Due to small size, worthwhile only for pets or show.

Emden: A pure white, erect goose (Fig. 1.11). It is a reasonably good layer (35 to 40 eggs per year) and a good setter. It is also a good meat bird, as it grows rapidly and matures early.

Pilgrim: Pilgrims are good market birds, and they have the added distinction of being sex-linked (color-coded by sex), a decided advantage for those of us who are not terribly proficient at feeling a goose's sex organs to determine gender. The day-old male goslings are creamy white and the females gray. The mature ganders (males) are white with blue eyes; the geese (females) gray and white with dark hazel eyes.

Sebastopol: This lovely white goose with curly feathers is good only for show or pets.

Toulouse: Toulouse geese make good mothers and lay about 20 to 30 eggs per year. They are also quite good as meat birds. They have broad bodies and look even larger because of the fluffiness of their feathers. They are dark gray on the back and the breast and white on the abdomen. Their eyes are dark brown to hazel, the legs and toes are deep reddish-orange, and the bill is orange.

Geese can benefit from crossbreeding. A particularly good cross is a White Chinese gander with an Emden goose. The result is a rapidly growing market bird. Experiment!

PURCHASE

You can buy fertile eggs for hatching, day-old goslings, or mature stock for breeding. If you plan to go into breeding, get sexed birds or buy enough so that you're sure you have a gander (preferably two, in case of accidental loss), and you won't have to buy one when your geese are grown. Flocks of geese tend

FIG. 1.11: EXAMPLES OF GOOSE BREEDS

to be very tight-knit, so if you plan to introduce a new member, prepare yourself for a battle.

Hatching

In an incubator, the procedure is the same as for chickens, except that the hatching time for Canada and Egyptian geese is 35 days, and 29 to 31 days for the other breeds. Incubators are difficult to use because more humidity is needed than for hatching chickens. Have a pan of water on the floor of the incubator, and try dipping each egg in lukewarm water for half a minute each day. Hatchability begins to decrease after the first week, but eggs can be stored up to 2 weeks before incubation without losing complete viability.

A goose or a duck foster-mother hatcher (preferably a Muscovy) can handle 10 to 12 eggs. Broody hens must be free from lice and, because of the size of the eggs, an average hen can handle only four to six goose eggs. The eggs need not be sprinkled with water if the goose or duck doing the hatching has daily access to swimming water. When using chickens, or ducks or geese without access to water, dip the eggs in water daily as in incubation. Eggs must be turned daily, as they are too big for a hen to turn herself. Also, when a hen is used, goslings must be removed from her as they emerge, and returned when all of them have hatched. If she deserts them or wanders away, confine her with them.

Day-Old Goslings

The brooder can be the same as for chickens but allow ½ square foot per gosling at the start, and increase to 1 foot at the end of 2 weeks. Depending upon the weather, artificial brooding is needed for only a couple of days to two weeks. If the weather is warm and sunny, you can let the goslings out after a few days, but be very sure they don't get caught in a shower or exposed to heavy dew. If you don't have a chicken brooder, you don't need to build one as elaborate as for chickens. Because they are so hardy, goslings can be brooded in a box with a light or in the corner of a room or barn. Make sure that no predators such as rats or weasels can get to them. Goslings can become quite attached to the person

feeding them, so much so that as they mature the bond may be difficult to break. In our setting, they became something of a nuisance. They refused to stay on our pond, preferring to make themselves comfortable under our lawn chairs as we attempted to get a few moments of relaxation in the backyard after a busy day.

HOUSING

In most cases, housing for geese is minimal or none at all. Geese raised in warmer months need a simple shelter so that they can be confined at night safe from predators; if predators are not a concern, they may be left to find their own shelter. A more permanent shelter for breeding purposes can be the same as indicated for ducks. Allow at least 8 square feet per goose for the larger breeds. Don't worry about them in winter, as they love to frolic in the snow and may spend more time outside the pen than in it.

If you have a large pond that does not freeze completely, geese can remain outside all winter. Geese have a innate sense that makes them remain in open water during the night, thus minimizing the chance of a predator attack. Snapping turtles are hibernating at this time of year, so pose no threat.

Don't just plop your geese down anywhere. Geese will roam quite a bit, so it is a good idea to house them as far from your garden as feasible. They are messy with their droppings and graze very closely, so rotation of pastures (if fenced in) or a location out of barefooted humans' traffic patterns is recommended. If you are forced to fence your geese because of predators or angry neighbors, use a 3-foot fence. Geese will also tear the bark from young trees, so keep them from young fruit trees or other trees you don't want to lose, or wrap the trees with wire as high up as the geese can reach.

EQUIPMENT

Feeding and watering equipment is the same as for ducks. The range feeder (a trough in which feeds such as scratch or mash can be placed) is especially important for large groups of geese. The waterer should be deeper so that they can immerse their heads, but not so big that they can swim in it. Swimming troughs, like those for ducks, can be provided if there is no natural water. Do not let them swim before three or four weeks of age or chilling may result.

Your nest boxes can be located inside or outside the house, because more often than not, geese like to lay their eggs outdoors. If the boxes are outside, place them in some sort of natural shelter or build a rudimentary one to protect them from the elements. You will need one nest for every three birds, and it should measure 2 feet square. If possible, space the nests a distance apart from each other to discourage fighting.

FEED

Fresh water must be available at all times.

Geese should not be fed medicated chick starter. Try to find a duck or goose starter (20–22 percent protein) and feed that for the first 3 weeks. If you can't find any starter, feed them bread crumbs wet with milk or a mixture of half cornmeal and half bran or oats moistened with milk. If the weather is suitable, you can let them out to forage for greens. In inclement weather, supply them with greens in the brooder, along with their regular feed.

The rest of their feeding program is the same as for ducks, except that you can rely more heavily on feeding whole grains, and geese will forage a great deal more than ducks do. Mix whole grains with the mash or pelleted feed at a 50:50 ratio. If you wish, for greater economy, you can feed only whole grains: corn, oats, bran, or wheat. If you are not feeding commercial feed with grit, you should supply some grit to your flock if they don't get enough in their foraging.

Geese favor the more succulent, tender grasses like clover, eschewing old pasture and hard stems and even alfalfa. Geese can get by on almost no grain the first month (and they don't seem to eat very much anyway) and therefore can be raised for very little cost. If you raise your geese on forage alone, finishing will be important for the best quality meat. If you confine your birds to a run and they exhaust their greens, it is essential that you supply them with plenty each day.

Finishing

A month or 3 weeks before slaughter, gradually work your geese over to a diet of one-half to one-quarter pasture and the remainder fattening feed. This finishing feed can consist of whole corn, cornmeal, barley, or a cornmeal and mash (or pellet) mixture moistened with milk. What you use depends mostly on what is available and cheap—experiment! As with any animal, do not change their feed overnight, but introduce the new feed gradually.

Breeding Diet

If you are wintering geese for breeding (or if you have become attached and want them as pets), you can feed them as for breeding ducks or a diet consisting of 15 to 25 percent breeder feed and the rest good, tender hay (preferably legume—second cutting is best) or silage.

Supplementing Commercial Feed

Their foraging ability makes geese one of the most economical meat producers to raise. In addition, you can grow much of the feed in the programs outlined above. You can plant a field of clover to supply exceptional grazing for your flock. Allowing geese to forage and feeding them corn (or barley, oats, etc.) that you grow yourself should reduce feed cost to an absolute minimum: the cost of the seed itself. With very little effort, you could grow all the grain you need for half a dozen geese.

Geese have the reputation for being natural weeders. It is my observation that this unique ability has been overrated. I have spent many summer hours waiting for my geese to eat the weeds in my garden, and for some reason I always end up doing the weeding while the geese watch in the shade of a tree.

MANAGEMENT
Handling

Geese will follow you almost anywhere if you coax them with a little grain. If you want to pick them up, secure them as gently as possible by the base of the neck, reel them in, and hold the wings gently with the other arm. Unlike chickens, grabbing them by the legs can cause injury.

Predators

Geese are threatened by the same predators as other poultry, but are able to ward off the smaller and more timid animals. They may not be able to fight off large dogs, but they are convincing bluffers. Still, a good house to lock them in at night, and as a last resort, a pen, will do much to prevent losses.

Goslings are, of course, more prone to predator loss. Rats and weasels can wipe out a flock if they can get at them in the brooder. Make sure the brooder is predator-free.

BREEDING

Select geese for mating that are healthy and rapid growers. One gander will service about four geese. They are not necessarily monogamous as some people suggest (except the Canada goose, with which we are not really concerned). Once they establish mating partners, it is difficult to introduce new geese to the flock. Generally, the laying and meat birds tend to mate in pairs or trios (i.e., one gander to two or three geese), while the smaller breeds may mate one gander to four or five geese. The best practice is to have plenty of ganders and watch their breeding habits to see if any of the ganders are unneeded. If they are, butcher them.

Sexing, in all breeds except the Pilgrim, is not the easiest thing in the world. To determine sex (and this takes practice), place the bird on its back (pointing the tail away from you) and bend the tail over backward a bit. Massage the genital area to relax the muscles and then press around the area with the thumb and index finger to expose the sex organs. The goose's will be rounder and more prominent. Practice on birds whose sex you know, and in time, it will be routine. This also works with ducks.

The Chinese may begin laying in the winter, but most geese will begin to lay in February or March as the hours of daylight begin to lengthen. They can be brought into production at the age of 7 months by increasing their exposure to artificial light. If a good pelleted breeder feed (or mash) is available, begin feeding that a month before production. If no duck or goose breeder is available, chicken laying mash or pellets can be used. If you are feeding a home-prepared feed, be sure to supply calcium (in the form of ground limestone, oyster shell, etc.) for eggshell hardness. Gathering eggs frequently will help cut down on broodiness. Hatchability will increase after the first year of production and is at its peak between 2 and 5 years. Geese will lay until about 10 years of age (and will often live many years past this), and ganders are good breeders for about five years.

During the breeding season, geese tend to wander in search of a nesting ground if they are allowed to roam freely. In some settings this characteristic may be undesirable, so it is necessary to restrict their range in the early egg-laying season.

Finding goose eggs may present problems for the owner, but it never seems a problem for the local raccoon population, which apparently considers goose eggs a treat. It's unlikely that a goose can hide her eggs from the raccoons, so collect the eggs as they are found and keep them in a cool place until a proper brooding area can be constructed. In our experience, allowing geese to set on their eggs outside is doomed to failure. I often wonder how wild geese can raise their broods in the face of the ever-increasing predator population.

HEALTH

Geese are among the most disease-resistant of all livestock. Normal preventive measures should ensure your flock against losses due to disease. The Appendix lists common poultry diseases, should you have problems.

BUTCHERING

Geese are ready for butchering at about 14 to 16 weeks. You can butcher at 12 weeks if you feed them extra heavily, but then again you're not saving money (and in fact are probably spending more), just time, and who's in a rush? Also, at the earlier age you'll have to do battle with more pinfeathers, and that is certainly no delight. The best time to butcher is in the fall, when the geese reach market weight and the supply of forage is dwindling.

Slaughtering is carried out as with other poultry. Save the blood (there's enough of it with geese to make it worthwhile) and the offal for other stock. The feathers are the worst to remove of all the poultry, but scalding and coating with wax, then finishing the job with a blowtorch, should make an attractive carcass. Some people hang the gutted carcass for a few days to age; suit yourself.

Down

There may be some market for the down in your locality, or you can most certainly make use of it yourself for stuffing pillows, quilts, sleeping bags, or your own parkas. If you wish to save the down, don't use wax when plucking. Wash the feathers in lukewarm water and detergent or water and borax and washing soda. Rinse them out and pat dry, then hang in an airy place in a cheesecloth bag.

You can also pluck live mature geese each spring and "harvest" the down year after year. Only mature geese should be used, and there will be no pinfeathers. Restrain the goose and pluck the down parallel to the body and toward the tail. Do this only after the weather is settled in the spring, and do not pluck them bald. They get embarrassed. Wash and dry the feathers as you would after plucking dead geese.

Turkeys

"When God gave out brains . . ." the old saying goes. Well, that day the turkeys got tied up in traffic. Ditto when they gave out resistance to diseases. In modern parlance, a "turkey" is a worthless individual, an idiot, a dolt. These comparisons hold, however, only for domestic turkeys. Wild turkeys survive quite well on their own, out of the reach of man. It seems when we domesticate animals, "improve" them for our own purposes, we make them dumber.

Despite their apparent shortcomings, no one seems to find much fault with their taste. Turkey is such a mainstay of American living that it was included in the first meal served on the moon. But when it comes to the home-raised, freshly killed turkeys, there is nothing—repeat, nothing—like their taste.

Turkeys are not the easiest poultry to raise, but raising them is definitely within the capabilities of the average homesteader. They require little space and time, and with sanitary conditions they should be disease-free. Almost anyone who raises animals should consider raising a few turkeys over the summer—a couple for family use, and a couple for sale to pay for the ones they keep and even make a little profit. If ever your turkeys get you down, if their stupidity amazes you, and you wish to be rid of them on the spot, be humbled by the old saying: "The only thing stupider than a turkey is the one who raises them."

BREEDS

There are three breeds that you will be most likely to come in contact with: the Broad-Breasted Bronze, the Broad-Breasted White, and the Beltsville White. The Bronze is the breed everyone pictures when they think of turkeys. Their plumage is brown or black, and because of the problems in plucking a

FIG. 1.12: BROAD-BREASTED WHITE TURKEYS

dark-feathered bird (getting a clean-looking carcass), the White has pretty well replaced the bronze as a commercial bird. The Beltsvilles are smaller and make good broilers at 15–16 weeks and good medium-size roasters at 24 weeks of age. The larger birds (Whites and Bronzes) make good large roasters at 24–28 weeks. The large-type hens can be butchered at only 13 weeks as fryer-roasters and can be butchered at 20 weeks to make good medium-size roasters. The main difference between the Beltsvilles and the larger Whites and Bronzes is that the Beltsvilles do not convert feed quite so efficiently, but they are usually cheaper to buy as poults (young turkeys).

PURCHASE

You will be able to choose between eggs and poults. Breeding turkeys is beyond the scope of small operations, requiring artificial insemination. Turkey eggs can be incubated much the same as chickens' eggs, except that the hatching time is 28 days rather than 21. After hatching, or upon purchase of poults, place the turkeys in a brooder. Brooding is the same as for chickens.

Turkeys must never be housed with chickens, nor should they be kept in the pen or brooder that has housed chickens within the past 3 months, or where chickens have ranged within 3 years. These precautions are to prevent turkeys from contracting diseases. If you use a brooder or pen before the proper time has elapsed, disinfect it thoroughly.

Allow the poults twice as much space and headroom as you would for chickens. Turkeys are more sensitive to drafts and dampness, so take pains to avoid such conditions. Also, cover the litter with rough paper for a week to prevent litter eating. Litter is not needed if you raise them off the floor on wire or wooden slats. Weather permitting, they may be removed from the brooder in 8 weeks. Round corners in brooder pens with curved cardboard or hardware cloth, to prevent the birds piling up in corners and smothering.

Purchase your eggs or poults from reliable breeders who guarantee disease-free stock. While you can order turkeys by mail, they do not ship as well as chickens, so you should try to buy locally.

HOUSING

There are three types of housing in most common use: confinement, sunporch, and ranging.

Confinement

This is very popular because it protects the birds from most predators as well as severe weather conditions. Again, dryness is one of the primary concerns. Litter must be deep and dry, and more added as it becomes damp. The birds can also be raised off the floor on heavy hardware cloth or on boards that are slatted a quarter-inch apart. Roosts are not necessary, but if you want them, construct them as you would for chickens. Allow at least 1 foot of space per bird on the roosts. In confinement housing, allow 5 square feet of space per bird

FIG. 1.13: TURKEY HOUSE WITH SUNPORCH

if they are debeaked, as described in the chapter on chickens, and 7–8 square feet if they are not. The drawback of this system and the sunporch method is that the turkeys won't be able to make use of greens and other supplementary feed the way range-reared birds can.

Sunporch Method

This is really identical to the above with the addition of a wire- or slat-floored "sunporch" attached to the house (Fig. 1.13).

Ranging

This can be accomplished by letting the birds out of their confinement house and allowing them to free-range. If you have chickens, they must not mix. You can keep them apart by keeping them at different corners of the property, as they usually don't range too far from their houses. If they still get together, fence in a run for your turkeys. Avoid

ranging them on poorly drained land, as the wetness may contribute to disease problems. Lock up your flock at night so they are not threatened by bad weather or predators. In warm weather they can be outdoors at 8 weeks of age; in more severe weather wait 10–12 weeks or until the weather has settled.

FEED

An average turkey will consume about 75 pounds of feed from birth to butchering, but you can be consoled by the fact that it has an excellent (4:1) feed conversion ratio.

Poults

From birth to the age of 8 weeks, young turkeys need a 28 percent protein mash, fed free choice. Mash must be fed up to 4 weeks; thereafter, a pelleted feed may be provided. Supply grit early and at all other times when a whole grain is fed and the birds are confined and don't have access to natural grit. Water should always be available. Young birds may need to be taught to eat. Tempting food such as coarser-milled chicken scratch or shiny marbles will usually attract them to the food and get them started.

Growing

From 8 weeks on until slaughter, the turkeys require a 20 to 22 percent protein growing ration. This comes in mash or pellet form and should be fed free choice along with some grain (corn is best, but anything that is palatable will be fine), also on a free-choice basis. Up to 16 weeks, whole grains should be cracked, but can be fed whole thereafter. Greens can make up a large part of a turkey's diet. If you don't range your birds, supply them with fresh greens once or twice a day. Free-ranging birds receiving fresh air, sunshine, and plenty of greens will require a less complete feed (no vitamin supplements needed) than birds raised in confinement.

If you feed a lot of greens to your turkeys, it is advisable to finish birds before slaughter. For 2 weeks prior to slaughter, they need only a 16 percent protein ration. Corn is the best for finishing and should be supplied free choice.

Supplementing Commercial Feed

Greens, available to free-ranging birds or supplied to those in confinement, can supply up to 25 percent of nutritional needs of your turkeys. Good greens include Swiss chard, alfalfa, tender grasses, grain sprouts, rape, lettuce, and cabbage.

For the size flock most people will have, it would hardly be worth the trouble to assemble your own mash as you might with chickens. Even if you did make your own chicken mash, it wouldn't be interchangeable because of the lower protein content. You can, however, make good savings by growing some of the grains you feed during the growing period.

Surplus milk can be mixed with the mash, but make sure it is cleaned up quickly and does not sour. Although I have never tried it, you can try feeding turkeys high-quality food scraps (equal to Grade I foods; see chapter on pigs) much as you would chickens. Because they demand a higher protein ration than chickens or pigs, the scraps should be rich in high-protein food such as meats and cheeses. Be sure that the feed is fresh and the birds don't sling it around the cages. Spoiled feed will invite predators and disease. It is a good idea to feed such scraps only to birds on range for reasons of cleanliness.

MANAGEMENT

Routines

Since colder weather may be a problem in raising turkeys, a spring-to-fall program is recommended. Buy your poults as soon as the weather moderates in the spring. Buy a half dozen and figure on keeping some for yourself and selling the rest. You should have no trouble selling them and might even turn a bit of a profit. Hens bring a slightly higher price than toms.

Handling

Much the same as with chickens. If you plan to "herd" them into a corner and grab one, be careful as this may cause them to panic, stampede, and injure themselves.

Predators

Predators are much the same as for chickens. A well-constructed, predator-free cage that you can keep the flock in at night will be your best insurance.

BREEDING

Breeding turkeys is beyond the scope of most homesteaders (and this book). If you do want to try it, information is available in a good poultry book, such as Leonard Mercia's *Storey's Guide to Raising Poultry*.

HEALTH

Strict sanitation and keeping your birds dry and free from chills will go a long way toward keeping your flock disease-free. Do not let the turkeys mix with chickens, and thoroughly disinfect any equipment that has been used by chickens in the 3 months before using it for your turkeys.

Medicated turkey feeds (which includes most, if not all, commercial feeds), if you're not against using them, should be quite helpful in preventing the two most common diseases, blackhead and coccidiosis.

The Poultry Disease Chart (see Appendix B) should be helpful in spotting and controlling most common diseases.

BUTCHERING

Butchering is the same as for chickens, except that the oil sac on the back near the tail should be removed, as it can cause a peculiar flavor.

CHAPTER 2

Rabbits

One 10-pound doe can produce in her litters up to 120 pounds of meat per year—a production of more than 1,000 percent of her body weight! To give you an idea of what this means, consider a good sow weaning two litters of ten 25-pound pigs per year. This would be a production of about 100 percent of her body weight. In order to equal a doe's output she would have to wean 200 such piglets, or 5,000 pounds of piglets!

Rabbits make an excellent backyard venture. They do not require a great deal of care, and they are efficient converters of feed to meat. The litters are butchered before weaning, so there is no need to go to the additional time and expense of building more hutches. If you have labored over the job of plucking chickens (and inhaling stray feathers), butchering rabbits will be a welcome change. A litter of eight to ten rabbits can be dressed in an hour or 2. The meat is similar, though I think superior, to chicken. It is all white meat and has a more delicate flavor. It exceeds the protein content of beef, pork, lamb, and chicken, but has a lower percentage of fat, cholesterol, and, ounce for ounce, fewer calories. The finished product is ready even sooner than veal, and because of the small bones, there is less waste (only 20 percent) in butchering than any other animal we cover in this book. As the world food picture grows gloomier, the rabbit may emerge as the animal of the future.

BREEDS

There are more than 100 breeds and varieties of rabbits in this country, but we need only be familiar with a small number. Rabbits can range in mature weight from 2 to 3 pounds for a Polish, to up to 20

Table 2.1: Weight Classes of Rabbits and Common Breeds Within Each

GIANTS	MEDIUM WEIGHT	SMALL	DWARF
Flemish	New Zealands	Tans	Polish
Checkered Giant	Satin	Dutch	Netherland
Chinchilla Giant	Champagne D'Argent	English Spot	Himalayan
	Californians	Havana	
	Chinchilla		

pounds for a Flemish Giant. There are four weight classes of rabbits, as shown in Table 2.1.

While one might think the giant breeds would be the logical choice for meat breeding, they are too heavy-boned and thick-skinned and too poor in feed conversion to be economical for meat use. The medium-weight breeds are light-boned and thin-skinned, so less feed goes into the production of skin and bones and more into meat (Fig. 2.1). Most breeds within this class produce rabbit fryers of 3 to 4 pounds in 8 weeks. Small and dwarf rabbit breeds are most often used for show and laboratory purposes.

The hides of medium-weight breeds, although not as high in quality as those of the fur breeds, are nonetheless suitable for home use. Your choice of medium-weight breeds will probably be limited because in most parts of the country rabbits have yet to be accepted on a large scale as meat producers. The most popular and available rabbits are the New Zealand Whites. They are hardy and fast gainers and therefore among the most desirable.

PURCHASE

If, after you decide on a breed, you can't locate a local source, contact your county extension agent or the American Rabbit Breeders Association (www.arba.net) for sources of your choice in your area. You might also check agricultural market bulletins or look for rabbit fanciers' clubs. If you still cannot locate a breeder, check in the back of the rabbit publications listed in the Appendix for mail-order services. This is more expensive and your chances of loss are greater, but it will give you a wider selection of breeds.

Age

Since rabbits don't reach maturity until 5 or 6 months of age, if you buy weaned rabbits (8 weeks old or so), you'll have to wait before you can begin

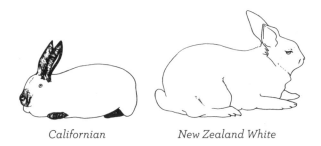

Californian   New Zealand White

FIG. 2.1: EXAMPLES OF RABBIT

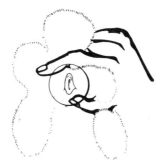

FIG. 2.2: SEXING RABBITS

breeding. On the plus side, they are cheaper than adult rabbits, and by buying young stock you will ease into the rabbit business, giving yourself extra time to iron out any problems before you have to become concerned about breeding. Buy young stock as early in the spring as the weather will permit, so they will be old enough to breed in the summer. Cold-weather breeding can be a problem, especially for first litters.

If you are ready to go into breeding, try to buy some 5- or 6-month-old stock, but be prepared to pay a slightly higher price. The age of rabbits is difficult to determine accurately, so you'll have to rely on the word of the seller (or try to get a look at a particular rabbit's hutch card if there is one). You won't run the risk of an over-age animal if you buy from a friend or a reliable dealer. Older rabbits should be avoided, because if an older doe has not been mated for some time she will often be sterile. Your older bargain-basement doe may never produce a litter. Old rabbits can often be spotted by their long and heavy toenails.

Start with one buck and two or more does. If you plan to have more than 10 does, you will need two bucks, as one buck is only able to properly service up to 10 does.

Sexing

Determining the sex of a young rabbit is difficult for the novice. To sex a rabbit, place it on its back and use the index finger of one hand to hold the tail out of the way. Using the thumb and index finger of the other hand, firmly but gently press down and manipulate the sex organ so that you expose the reddish mucous membrane (Fig. 2.2). In bucks, the membrane will protrude to form a circle, while in does, it will form a slit with a small depression toward the rear. An easier, although less reliable, method of sexing is to hold the rabbit by the scruff of the neck and feel along the belly. If the area is smooth, it is a doe; if it's bumpy, it's a buck. In older rabbits, the testicles will be prominent in males, so sexing presents no problem. With enough practice, you will be able to sex a rabbit quite easily. In the manipulative technique, do not handle one rabbit too long as irritation may result.

In selecting breeding rabbits for meat, look for specimens that are chunky and blocky as opposed to the long, rangy ones. They should have large feet without any sores on them. Similarly, the ears should be free from sores, and the mouth should be well-formed and not have teeth that are broken, or "buck teeth" (malocclusion). The rabbit should appear alert and have bright eyes and a smooth, glossy coat. Inquire about the rabbit's parentage, as their characteristics were passed on to the rabbit and will in turn be passed on to its offspring. All large rabbitries have extensive hutch records that you can look at to help you in your choice. (This is another good reason to buy from an established rabbitry.)

Look for stock that comes from large litters whose mother consistently weaned seven to ten young. Few deaths prior to weaning would also indicate a good milk supply as well as disease resistance. Studying a hutch card will also give speed of growth as well as the number of litters per year and other pertinent data. Remember, as with other stock, it is better to settle for the worst from a good litter than the best from a poor one.

HOUSING

Rabbit hutches should be placed where the rabbits will be protected from rain and snow, windy conditions, and neighborhood dogs and children.

Rabbits do not require a great deal of sunlight, so they can be raised in a barn or garage. If they are kept outside, they should not be exposed to too much sunlight, particularly in warm weather. Sunlight helps keep the sanitation at a satisfactory level, but whether it helps the rabbits hasn't been determined.

The flooring under the hutches should be concrete or dirt. A wood floor will absorb urine, thus creating an unsanitary condition. A concrete floor requires frequent cleaning. Wood chips, sawdust, or peat moss under the hutches will absorb moisture and minimize odors.

After years of trial and error with various materials, commercial growers have settled on all-wire hutches. They are inexpensive to build or buy, lightweight, will last for years, require little cleaning, and provide a healthy environment for the rabbits. Cages built with wood and wire absorb urine and are gnawed by the rabbits.

Farm supply stores stock these wire cages. But if you plan to have a dozen or more breeding rabbits (and you need a cage for each one), it is far cheaper to build your own hutches. Even the person with 10 thumbs can build very satisfactory ones.

The following instructions are for building a hutch that is 36" x 30" x 18" high. This is large enough for a medium-sized rabbit and her young.

The Wire Hutch
Tools:
Tape measure or yardstick
Hammer
Pliers

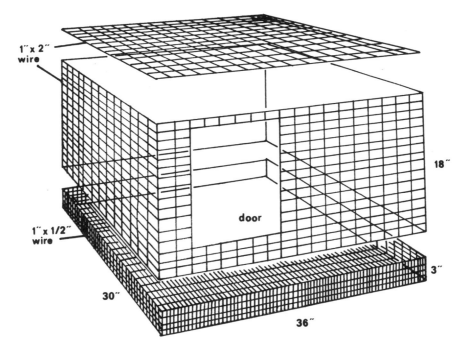

FIG. 2.3: BUILDING THE RABBIT HUTCH. *This hutch is built with three pieces of fencing: the top, the four sides, and the bottom.*

Wire cutters
J-clip pliers
One piece of 2 x 4 or 2 x 6 board, about 2' long

Materials:
One piece 14 gauge 1" x 2" welded wire fencing,
    18" wide, 11' long
(four sides of hutch)
One piece 14 gauge 1" x 2" fencing, 30" wide,
    36" long (top)
One piece 14 gauge 1" x 2" fencing, 12" x 14" (door)
One piece 14 gauge 1" x ½" fencing, 36" wide x
    42" long (bottom)

80 J-clips
Door latch

Forming the sides: First, take the natural curl out of the 11-foot-long piece and the top pieces of fencing. Place each piece on the floor so that the center, not the ends, tends to rise. Place the 2 x 4 or 2 x 6 on one end, stand on it, then gently pull the other end of the fencing toward you enough to remove the curl.

Turn the side piece over. Now, as you bend it up to make the corners, the bend will not tend to strain the weld but will push against it.

Measure 36 inches from one end, place the piece

To form 3-inch sides extending up from the bottom, first cut 3-inch squares from each corner of the bottom piece.

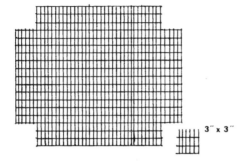

3" x 3"

To form corners, hold wire pieces against the board, then tap individual wires with the hammer.

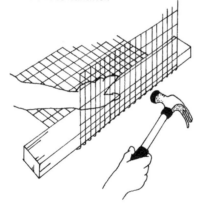

J-clips hold wire sections together.

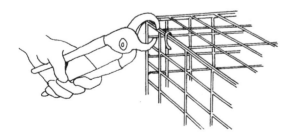

of wood at that point, lift the 36-inch section up against the 2 x 4, then tap each wire with the hammer to form a 90° angle. Continue to create, in turn, the 30-inch, 36-inch, and 30-inch remaining sides. Using the J-clip pliers and J-clips, fasten the two ends to form the box, placing J-clips every 4 inches.

Forming the Bottom: Do not flatten this 36" x 42" bottom piece unless it is very tightly rolled. Its natural curl will help to prevent the floor from sagging.

This bottom piece is cut large enough to provide both a bottom and four 3-inch sides of this tighter wiring so tiny rabbits won't fall through and out of the hutch.

Cut a 3-inch square from each corner of this piece. Check to see that the half-inch wires are on top, where the rabbits will be standing; then, using the hammer and piece of 2 x 4, bend the 3-inch sides up. Fasten this bottom to the sides, using J-clips at the corners and again every 4 inches.

Putting on the top: Put the top in place and fasten it with J-clips every 4 inches.

Cutting the door: If you're right-handed, you'll find it more convenient to have the door 4 inches in from the right side of the 36-inch front of the hutch. If you're left-handed, place it on the left side. In either case, cut it 4 inches from the bottom, and make the doorway 12 inches wide and 11 inches high.

Don't cut the wires flush. Leave a stub, then use

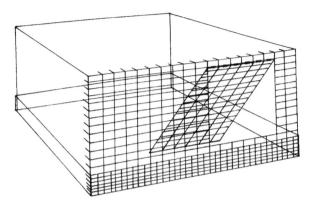

For safety's sake, hang door so that it swings inside and up.

the pliers to bend the stubs around the adjoining wire. That way, there's no sharp projection to scratch the rabbit, or your arm.

Installing the door: A latch will be needed. Small latches made especially for wire doors can be purchased, or you can improvise with a dog leash clasp, binder ring, or even an oversized paperclip. If you purchased a latch, install it now.

Install the door centered over the doorway space, hanging so that it will swing inside and up. This is the better way to hang it for several reasons, among them that if you forget to latch it, the resident rabbit may not discover your error. Use five J-clips, leaving them quite loose so the door will swing easily.

For convenience, you may want to fashion a small hook from coat-hanger wire and hang it from the top of the cage to hold the door open.

Placement of the hutch: This hutch can be used in several ways. It can be placed on a dirt floor, raised on four wooden legs (2 x 2s are fine), suspended by light chains from beams, or, as many commercial growers do, mounted two deep on an angle iron frame built to hold a dozen or more hutches.

If you find that it is necessary to have one hutch mounted over another, the lower cage should be shielded from the one above by a sloping piece of light galvanized sheet metal between the two, to avoid contamination from urine and feces.

This hutch is particularly well-designed for the rabbit raiser who is going to build a large number of hutches and can take advantage of buying fencing at the cheapest rate, in complete rolls of the required widths. In estimating your needs for several hutches, remember that 2-feet of fencing are lost each time you cut the 1" x 2" of fencing.

EQUIPMENT

The hutch we have just discussed will provide clean and comfortable conditions for the doe and her young and is ideal for the person who wants to minimize the work involved in feeding and watering the rabbits. Careful planning and the right equipment can cut the time for this daily chore to only a few minutes.

Feeders

Ideally, you will buy self-feeders. They are not expensive and will return the initial cost in saving feed that otherwise would be wasted. The feeder is mounted on the front of the hutch so that the trough extends through a hole in the fencing into the hutch while the hopper is on the outside, where it is easily filled.

Similarly, a hay feeder can be built so that hay can be inserted from the outside and is within the reach of the rabbit. Use pieces of 1" x 2" fencing to build a box with an 8" x 8" bottom, 6" x 8" sides, and a 6" x 8" back. Using J-clips, build this box, then attach it to the top, with the two sides attached to the front of the hutch. An ideal location is centered above a self-feeder. Cut the wire in the front to provide a way of putting hay into the feeder. Bend stubs around the adjoining wire to avoid scratches.

This hay rack has an advantage over hay racks mounted on the side of the hutch. Any hay dropping out will fall into the hutch and be eaten by the rabbit, rather than falling on the floor outside and being wasted.

Waterers

Keeping a supply of water within reach of your rabbits is a must. Many who keep rabbits use cans or crocks as containers, but these have the obvious disadvantage of fouling by the rabbits, or spillage when the rabbits tip them over.

One alternative is to purchase a bottle waterer with a drinker tube. This is fastened to the outside of the cage with hooks and wire, and the tube extends into the hutch.

You can make a very satisfactory waterer using

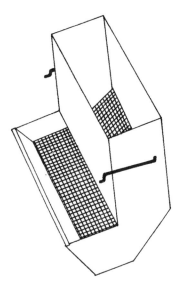

FIG. 2.4: *A self-feeder that can be filled from outside and that has a lip inside that keeps rabbit from scratching pellets out and thus wasting them.*

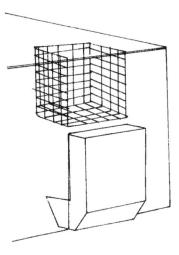

FIG. 2.5: *Hay rack that lets rabbit pull hay down into hutch.*

the largest size plastic soft drink bottle. Buy a drinker valve (used for automatic watering systems) and mount the valve near the bottom of the bottle by cutting a hole for it, then setting it firmly in place with epoxy cement. Hold it on the outside of the hutch with wires placed so they can be slipped off when the bottle must be refilled. Position the bottle so the valve extends into the hutch.

Both semiautomatic and automatic systems can be installed, using the drinker valves and black plastic pipe. The semiautomatic system requires a tank placed high so that the system is gravity-fed.

The automatic system is linked to a piped water supply. It requires some method of reducing the water pressure before the water enters the plastic pipe. One method is to pipe water from a tank, with the intake to the tank controlled by a float valve such as is found in a toilet tank.

Both of these systems will freeze in cold weather unless protected. One method is to drain the entire system, a tiresome job if it has to be repeated many nights. The other, more satisfactory method is to insert electrical heating cable inside the pipes.

Nest Boxes

Nest boxes are essential when the young are born and until they are strong enough to venture into the hutch.

The first rule of building is to make sure a nest box is small enough to fit through the 12" x 11" hutch door since it must be put in for each litter and taken out when the little rabbits no longer need it. A wood

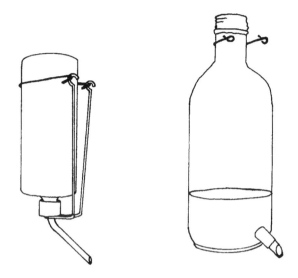

FIG. 2.6: THE TUBE BOTTLE WATERER (LEFT) AND BOTTLE WITH DRINKING VALVE (RIGHT)

box 11" x 18" x 8" high can be built. To avoid having the rabbits gnaw the wood, all edges must be covered with metal. Moisture holes should be drilled in the bottom.

Better than the wood box is a wire box of the same dimensions, but with the 8-inch front cut down to only 4-inches high. Make this of 1" x ½" galvanized wire fencing.

Cut one 18" x 27" piece for the bottom and two sides, one 11" x 8" piece for the back, and a final 11" x 4" piece for the front. Bend the sides and bottom using the method explained in the construction of the hutch. Attach the pieces with J-clips at all corners, 4 inches apart.

The four top edges of the box should be covered

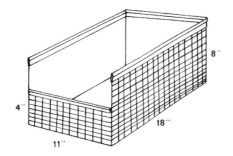

FIG. 2.7: *A nest box large enough for a doe and her young and small enough to fit through hutch door.*
Shaping the metal strip around the dowel produces a keyhole shape for the flange.

with flanges to avoid injury to the youngsters. Make these of 3-inch wide pieces of galvanized sheet metal of the correct length. Bend the strips around a ½-inch wooden dowel, then hammer the metal to form a small circle atop two ends. Fit these flanges over the top edges of the back, sides, and front.

Complete this box by cutting a piece of heavy cardboard the size of the bottom. Poke drainage holes through it. Place it in the bottom of the box, then cut a piece of the fencing the same size and place this over the cardboard. This will keep the mother rabbit from chewing down through the cardboard.

Place several inches of shavings plus lots of straw in the box. These are nesting materials for the doe, who will add to the pile by pulling fur from her own body.

In cold weather, insulate the four sides of the box using pieces of cardboard. This cardboard and all the other nonmetallic material in the box should be replaced after every litter.

Miscellaneous Equipment

A salt spool, available at feed stores, should be furnished in each cage. Also, a small block of wood thrown into the cage for the doe to chew on will help to keep her teeth in proper condition, and may help reduce chewing on any exposed wood.

It is recommended that you keep complete records for all your livestock. and for rabbits, a hutch card. The card, located either on the hutch or in a record book, should list such important information as the tattoo number of the animal, date bred, service buck ID, kindling date, number born, number weaned, weight, and other data. On a buck's card, include all pertinent information on breeding and his offspring. These records will enable you to select good breeding stock and cull poor producers. (Such cards are usually available in feed stores.)

Tattooing is the preferred method of identifying rabbits. Ear tags and clips are not suitable as they invariably tear out. Tattoos and tattoo pliers are available from most large livestock supply houses.

FEEDS AND FEEDING

The digestive tract of the rabbit differs from some of the other animals under consideration in this book in that, in one sense, it is rather unique. First, let it be said that rabbits need fiber in their diet, but digest it rather poorly.

Fiber is needed in the diet to ensure that the contents of the intestinal tract move along at the proper rate. The cecum is a blind pouch located at the end of the small intestine, an organ that is comparable to our appendix except that in the rabbit it fills a very real need. You will remember that in cattle, much of the fermentation of feed takes place in the rumen. In the rabbit, a somewhat similar fermentation process takes place in the cecum. It is here that many of the smaller particles undergo fermentation and are eventually passed as soft feces which are consumed directly from the anus. It would be unusual to see this, because most of this feeding takes place at night. This material contains several of the essential nutrients required by the rabbit including B-complex vitamins and vitamin K, fatty acids, and some protein. The hard pellets that we associate with rabbits are primarily breakdown products of fiber that bypass the cecum and move directly to the outside. A low amount of fiber in the diet can lead to a variety of problems, the most serious of which is enterotoxemia, which may result from a low rate of passage of material through the bowels. This low level gives certain bacteria that reside in the bowel the opportunity to grow, producing a lethal toxin, especially if the fiber level in the diet is too low.

Commercial rabbit pellets are higher in protein (18 percent to 20 percent) than most other feeds, and are, on a per-pound basis, more expensive than most others. However, the rabbits' high feed-to-meat conversion ratio (3.5:1) makes it possible to raise meat economically using only commercial feed.

Almost as many problems arise from overfeeding stock as when they are underfed. It is rather difficult to set any hard and fast rules for feeding rabbits, since their nutritional requirements vary with breed, size, age, and weather conditions. Generally speaking, a mature rabbit that is healthy and not pregnant will consume 3.8 percent of its body weight in feed daily. Therefore a 10-pound doe will eat about 6 ounces of feed per day (10 x .038, or .38 pound). Because of the variables mentioned above, this calculation should serve only as a guideline so that beginners don't give a rabbit 10 pounds of food a day or expect it to subsist on three pellets.

As you learn more about rabbits and your own stock, you will be able to tell if they're getting the correct amount. If possible, have an experienced breeder show you a properly maintained animal. Get the feel of it, especially around the ribs and backbone. If a knowledgeable rabbit person is not available for advice, pick out one of your own rabbits that is producing well and study it. If the flesh over the backbone and ribs is spare, increase the feed. If there is too much cover, reduce the feed intake and continue to make adjustments until you reach the optimal level. In time, and with daily handling of your herd, you will be able to maintain your rabbits well and spot and correct any problems quickly. Oftentimes there is more art than science involved when it comes to animal feeding and also the innate instincts of the person doing the feeding.

There are two varieties of pelleted rabbit feed: an all-grain pellet designed to be fed with supplemental hay and complete pellets containing all the nutrients necessary for a completely balanced diet.

Junior does and bucks, mature dry does, and herd bucks not in service but in good physical condition can all be maintained on hay alone if it is a fine-stemmed leafy legume. If coarse grass hay is the only type available, it can be fed if each 8-pound animal is supplemented with 2 ounces of an all-grain pellet several times a week. Adjust the amount of grain for rabbits of varying weights. For example, a 4-pound rabbit would require 1 ounce of pellets, while a 12-pound animal would require 3 ounces.

Alfalfa pellets may be fed to developing junior does and bucks as their only feed from the time they are weaned until they are ready for breeding. Buy pellets containing 1 percent salt and 99 percent of No. 2 leafy or better-grade alfalfa meal (15 percent protein).

Feeding Does/Creep Feeding

In feeding pelleted feed to pregnant does and does with litters, allow them all they will clean up between feedings. Creep feeding is placing feed in a container accessible only to the young rabbits prior to weaning. While not essential, this method will ensure more rapid and more economical gains (see

Chapter 3, Fig. 3.3 for lambs). The doe's milk is the ideal food for young rabbits (Mother Nature's grand design!) and is all that is needed until slaughter. Its high protein content (15 percent) makes it one of the richest milks of all breeds of domestic animals. If you want to experiment with creep feeding, supply pellets free choice from a creep feeder. Again, as with all animals, experiment; compare weights of those litters that are creep-fed and those that are not, and decide whether the added cost of feed is worth it.

You can feed your rabbits as many times a day as you want, but one feeding is adequate, as long as it's enough. *Be consistent.* Rabbits eat the most in the evening and night, so if only one feeding is supplied, give it in the evening.

Contrary to popular belief (reinforced by cartoons of Bugs Bunny subsisting only on carrots), rabbits cannot live exclusively on carrots and greens. If you feed greens (see below), allow your rabbits only what they will clean up in 10 to 15 minutes in the evening.

Water/Salt

As mentioned earlier, a mother and her litter must have water available at all times, for they may consume up to 1 gallon of water in a day. Fresh water should always be available. Salt, in the form of a salt spool, should also be provided free choice.

Supplementing Commercial Feed

In conventional feeding—providing only commercial feed—you have a complete food and won't have to be

Table 2.2: Feed Requirements for Does	
	PERCENT RATION (BY WEIGHT)
Protein	20
Grain	39.5
Roughage	40
Salt	0.5
TOTAL	100

Table 2.3: Feed Requirements for Dry Does and Bucks	
	PERCENT RATION
Protein	8
Grain	31.5
Roughage	60
Salt	0.5
TOTAL	100

concerned about other nutritional needs. In supplementary feeding, you must have some knowledge of rabbits' nutritional requirements to be certain that they are met. In simplest terms, does (pregnant, or with litters) need a 20 percent protein ration. A pelleted feed supplies this, or it can be provided in the formula given in Table 2.2.

Dry does and bucks need about an 8 percent protein ration. They can, of course, be fed commercial feed, but they are getting much more protein than they need (20 percent). It won't hurt them, but is not cost efficient. You can feed other things, as we shall see, but the general formula is given in Table 2.3.

Commodities that make up each of the above groups are given below. Some of these you might be able to grow, or come by cheaply; still others you will have to purchase commercially. In any event,

making use of the above formulas and experimenting with combinations will enable you to decide on a food mix that is the best for your animals and is the most economical for you.

Sources:

Protein: In order of palatability and nutritional value for rabbits are peanut meal, soybean meal, linseed oil meal, hempseed meal, and cottonseed meal. Whole soybeans, while high in protein, are often less palatable for rabbits. Pellet meals or pea-sized cakes of the above meals are best.

Grains: Cereal grains in order of desirability are oats, wheat, and grain sorghums, barley, corn, and milo. Barley and oats need not be rolled. Bread or other wheat products can be used.

Roughage: Any good legume hay (second cutting is best). It is eaten most efficiently if cut into 3- or 4-inch pieces. Carbonaceous hays such as Sudan grass and timothy can be fed, but the protein content of the feed should be increased a bit, as those grasses are somewhat lower in protein than legume hay.

Salt: Salt can be mixed in with the feed at the indicated rates. Better, supply a mineralized salt spool for all your rabbits.

The Walters family of Campbellsport, Wisconsin has developed the feed program given in Table 2.4 and has reported good success with it. It may not perfectly suit your rabbits, but it can be used as a starting point in formulating a more self-sufficient ration for your rabbits.

Sally Cook of Rochester, New Hampshire reports success with raising rabbits on a program that includes millet, squash, and apples. She grows a plot of millet that she feeds to her herd unthreshed and unwinnowed, allowing them to eat the hay along with the grain, which the rabbits husk themselves. She collects culled or damaged squash from her garden and from a nearby squash farm for winter feeding. She allows her rabbits as much of it as they will clean up in a day. They will eat the pulp, but enjoy digging for the high-protein seeds inside. Apples that are picked in abandoned orchards and are left over from her cider-making are fed during the winter. While she does not gain top efficiency and has only five or six litters per year, she maintains that her herd is healthy and provides her with good meat at a minimum cost.

Table 2.4: Walters' Feed Program

SUMMER
DRY DOES AND BUCKS:

Morning and evening: Greens and weeds from garden (see Table 2.6 on Desirable and Undesirable Greens) No more than they will clean up in 10–15 minutes.
Evening: Grain supplement (4–6 oz. of raw grain consisting of 3–5 oz. oats and 1–2 oz. soybeans)

PREGNANT DOES AND DOES WITH LITTERS:

Morning and Evening: Weeds and greens (see above)
Evening: 2–2-½ cups grain mixture consisting of 30 percent corn (omit in warmer months), 60 percent oats, and 10 percent soybeans.

FALL AND WINTER

Hay and a grain mixture of 45 percent corn, 45 percent oats, 10 percent soybeans. Feed free choice to active does, and enough to maintain dry does and bucks.

Other rabbit feed recipes that can be made up by the industrious homesteader, as supplied by the US Department of Agriculture, are given in Table 2.5.

To make a feed mix, assemble all the ingredients in the correct proportions and feed it to your rabbits. If they get picky and eat only "favorite" parts, have the ration ground. Whenever you feed mash to rabbits, be sure to wet it down a bit, as dusty mash will interfere with a rabbit's breathing.

Since rabbits eat very little compared to pigs, any food surpluses you use will result in substantial savings. If you can grow corn and have access to stale bread, you have two food groups accounted for (using the food program in Table. 2.4) and need buy only a protein supplement to complete your ration. Often farmers will, for the asking, allow you to glean their fields after harvest in the fall. People I know collect left-behind ears and for a few days' work, have all the corn they need for a winter.

To get corn off the cob, you will need a corn sheller that is available from feed stores or livestock supply catalogs. Oats need not be rolled, as rabbits will shell them themselves, but rolling will prevent waste. For rabbits that will eat it, the chaff can make good roughage. There may be more waste when feeding whole grains, so it's a good idea to place a board under the feed dish to collect dropped food.

Vitamin supplements to supplementary feed are generally unnecessary, because the vitamins A, D, E, and B should occur naturally in all feeding routines.

Vitamin A is present in some root crops (carrots, Jerusalem artichokes) and good hays. Vitamin D is present in hay that is sun-dried, and vitamin E is present in dry hay, grains, and protein supplements. Vitamin B is synthesized in the stomach of the rabbit in a type of pseudo-rumination, and the vitamin is obtained by eating a small amount of mucous-coated feces that is passed at night.

As stated earlier, when fed with care, greens can supplement feed. Young rabbits should never be fed greens, as they may fill up on them and ignore more nutritious food. In limited amounts greens are quite helpful, but excessive intake should be avoided because of their high water content. Table 2.6 lists good and bad greens.

Table 2.5: Rabbit Feed Recipes

FOR PREGNANT DOES WITH LITTERS

(1)		
Whole oats or wheat	15 percent
Whole barley, milo, or other grain sorghum	15
Soybean or peanut meal pellets or pea-sized cakes (38–43% protein)	20
Alfalfa, clover, or pea hay	49.5
Salt	0.5
	100

(2)		
Whole barley or oats	35 percent
Soybean or peanut meal pellets or cakes	15
Alfalfa or clover hay	49.5
Salt	0.5
	100

(3)		
Whole oats	45 percent
Linseed meal pellets or cakes (38–43% protein)	25
Carbonaceous hay	29.5
Salt	0.5
	100

FOR DRY DOES, BUCKS, AND YOUNG RABBITS

(1)		
Whole oats or wheat	15 percent
Barley, milo, or other grain sorghum	15
Legume hay	69.5
Salt	0.5
	100

(2)		
Whole barley or oats	35 percent
Legume hay	64.5
Salt	0.5
	100

(3)		
Whole oats	45 percent
Soybean, peanut, or linseed pellets or pea-sized cakes	15
Carbonaceous hay	39.5
Salt	0.5
	100

MANAGEMENT
Routines
While not too many Americans are familiar with rabbit meat, once they have tasted it they often become converts. Invite some friends over for a rabbit dinner, and I'll bet they will become a ready market for your surplus meat rabbits (or for some weanlings, as breeding stock) and get started themselves. By selling your surplus meat, you may be able to pay for your own meat and even turn a little profit.

As an added plus, rabbit manure is one of the most valuable kinds. You can use it for your own flowers or vegetables or find a market for it with earthworm farmers—or maybe start growing your own earthworms. The pelts, while not valuable commercially, can be put to many uses around the house for gloves, hats, and blankets. Do not be tempted to sell little bunnies as Easter gifts for children. While it may be seen as a source of added cash, more often than not it also ends up in death or serious mauling for the rabbits. Rabbits given as Easter gifts probably are the most ill-treated of all pets.

Chewing: To prevent chewing and the consequent destruction of the wood portions of your hutches, cover any exposed wood with tin or wire. Blocks of wood and twigs will help to satisfy the rabbits' urge to chew and thus condition their teeth. (Remember what Bugs Bunny looked like before he went to the orthodontist.)

Molting: Chickens lose their feathers; rabbits lose their fur. This shedding and restoring of the coat normally takes place once a year in mature rabbits and in the first several months of age in young rabbits. It may be caused unnaturally, by disease, sudden high temperatures, rapid changes in diet, or other stressful situations. Normally, molting should be a minimal concern except that extra feed will probably be needed, and conception in does and production of sperm in bucks may be adversely affected.

Handling
Carrying rabbits around by their ears will result in injured animals. It should be left to magicians, who can presumably make their injured animals disappear. The correct way to pick up and carry a rabbit is by grabbing the skin of one shoulder in one hand and placing the other hand under the rump for support. The feet should always be pointed away from

Table 2.6: Desirable and Undesirable Green Feeds and Root Crops for Rabbits

DESIRABLE

Rapidly growing cereal grains	Jerusalem artichokes
Rape	Green hay crops
Trimmings from leafy garden vegetables	Kale
Dandelions	Lawn grasses
Sweet potatoes and vines	Tender twigs
Plantains	Filaree, alfilaria, or stork's bill
Mangels	Malva or mallow, cheeseweed
Swedes	Carrots and tops
Kohlrabi	Turnips
Culled potatoes and peelings	Sugar beets

UNDESIRABLE (TOXIC AND TO BE AVOIDED)

Burdock	Miner's lettuce
Castor bean	Nightshade
Fireweed	Oleander
Goldenrod	Poppies
Horehound	Sweet clover
Lupine	Tarweed
Milkweed	Rhubarb

Source: From *Domestic Rabbit Production* by George S. Templeton. Danville, IL: Interstate Printers & Publishers, Inc., 1968, p. 59.

you to prevent scratching. For an animal usually regarded as placid, the kick of a rabbit to the unsuspecting handler can result in a rather nasty wound.

Predators

Cats, dogs, opossums, weasels, snakes, and rats all share with humans a love of rabbit meat. Housing your rabbits in a barn or other closed building will eliminate the most common predator, dogs. Well-constructed cages using wire should keep out most predators. Safeguard your young rabbits by having a thick bottom on nest boxes to prevent rats and other predators from gnawing through and getting at them. If rats or snakes climb up the legs of your pen and into the cages, you should put circular sheets of metal around each leg to block their path.

BREEDING

The age of maturity varies with the weight class of the rabbit. Medium-weight rabbits usually reach sexual maturity by 6 months but sometimes as late as 7. The dwarf breeds mature at 5 months, and in the giant breeds maturity is reached from the tenth to the twelfth month.

The doe's eggs (ova) are contained within small follicles, cyst-like structures located on the ovaries. Nine to 10 hours after mating, the follicles rupture, releasing the eggs. This is known as ovulation. The sperm that was deposited at the time of mating will fertilize egg cells within 2 to 4 hours after the follicle breaks.

The doe should be taken to the buck's cage for

mating. This will tend to prevent fighting, as a doe is liable to attack a buck if he is taken to her cage. He will usually mount her without incident and, when mating is finished, he will often grunt and roll over. There is some controversy about whether conception rates are improved by returning the doe for re-mating in a few hours. This is a unnecessary procedure. Viable sperm are well on the way up the fallopian tubes to meet the recently released ova by the time a second mating would likely be made, and again Mother Nature has ensured fertilization by providing a huge number of sperm to meet the tumbling ova.

In some cases the doe will not accept the buck. If she ignores him or growls, do not leave them alone. They might fight and injure each other. Watch them for a while. If the doe evades him but her tail is twitching, she probably will accept him in time. If, however, the doe does not accept him, you might try moving her cage next to his for a few days so they can get used to each other. Often she will accept him when he is reintroduced. Another method which often succeeds is to switch cages. After a few days, take the doe back to her own cage, leaving the buck inside.

If you want results immediately, you can try force mating. Grasp the doe by the ears and the skin of the shoulder with one hand and place the other underneath her with two fingers on either side of the vulva, pushing it out a bit. As the buck mounts her, raise her hindquarters a bit to allow normal coupling. This will not ensure conception, as there are 4 days in the doe's 16-day cycle when no eggs are released for conception. During these 4 days she lacks interest in the buck, thus leading to rejection.

The gestation period is an average of 31 days—as short as 29 and as long as 35. Three to 5 days before kindling, or giving birth, place the nest box and bedding (straw or hay, but not sawdust as it can asphyxiate the young) in the hutch. Kindling is imminent when the doe begins plucking her fur to line the nest box.

The day after kindling, offer the mother rabbit a tempting morsel of food, then quietly remove the nest box and inspect the young. Remove any dead or deformed babies and any runts in a large litter. At this time you may even out litters by taking some babies from a large litter and letting a doe with a small litter nurse them. This can be done with litters which have kindled up to 3 days apart and is a good reason why you should breed your does to kindle about the same time. While litters may range in size from one to 20, eight to ten is an ideal number for a doe to handle.

At about 10 days of age, the kits' eyes will open and soon the babies will be in and out of the nest box. They will begin eating some of the doe's food and if you choose, they can be creep-fed.

At 8 to 10 weeks of age they should be weaned and their fates determined by the demands of the marketplace.

Problems

Your doe may kindle her young on the floor outside of the nest box, or refuse to nurse her young, or in extreme cases may kill them. Usually the cause is simple, such as moving her to a new cage too soon before kindling, excessive noise, or the presence of predators. Does are often quite sensitive close to kindling time, and it is wise to leave them alone as much as possible. If you find young on the floor of the cage, don't panic. When I was young, a doe of mine kindled eight on the floor of the cage, but upon close inspection I found that there were another ten in the nest box. I put them back in the nest box and she just as promptly returned them to the floor of the cage. Apparently, as is so often the way with animals, it was a case of not giving her credit for doing what nature told her. She was evening off the litter herself, as she apparently knew if she tried to nurse all 18, most would die, or at best be so weak that they would fall victim to disease.

If does fail to pull their fur off and then kindle on the hutch floors, make a comfortable nest and pull enough fur from the doe to cover the litter. Keep some extra fur on hand for such eventualities. Subnormal body temperature (hypothermia) is responsible for many fatalities among the very young, so supplemental heat is always a good idea. Many times it is possible to save litters that appear completely lifeless just by warming them.

Occasionally, the eyes of baby rabbits may become infected and fail to open normally. If these infections are treated promptly, the young rabbits generally recover without any permanent eye injuries. If the eyes appear to be infected, bathe them with warm water until the tissues soften and the lids can be separated with slight pressure. If a pussy discharge persists, an antibiotic eye ointment should be obtained from your veterinarian.

If a doe won't nurse and you find there is no milk, mastitis—a bacterial infection of the mammary glands—may be responsible. At onset, the udder will be pinkish and hot and the teats may be pink or light blue. Wash and disinfect all equipment and the nest box. For treatment to be effective, an appropriate antibiotic must be given at the recommended intervals to achieve a successful resolution. Death may result if treatment is not given promptly. Let her nurse her young, if she can or will. Do not resort to a foster mother as the infection may be spread to her; because any kits having already nursed on the infected nipples may harbor the bacteria in their mouths.

In the case of orphaned litters or when a doe cannot nurse, you can foster some on another doe if they were born within 3 days of her own. To prevent her from rejecting them, put some petroleum jelly on her nose. In cases where you have no other doe to pass orphans or "extras" from a large litter onto, you can raise them yourself but, please, think twice. Raising any animal from birth is a taxing job, and rabbits are among the most demanding. They must be fed with an eyedropper or doll bottle at least four times a day (and at night). If you're not sure you'll stick with it, dispose of them in as humane a manner as possible. This is preferable to nursing them for a few days, probably at a low level of nutrition, and then deciding the considerable effort is no longer worth it.

If you decide to raise them yourself, use cow's or goat's milk, or evaporated milk mixed with water to the consistency of whole milk. Another good formula is one half cup evaporated milk, one half cup water, one egg yolk, and one tablespoon Karo syrup. Heat the milk so it is warm on the back of your hand and feed it with an eyedropper *at least* four times a day (eight times a day is preferable for the first 4 days). At 2 weeks, they can be offered rolled oats and a few blades of grass. By 15 to 18 days, they can be taught to drink from a saucer and eat regular rabbit feed. Moistening the feed slightly will make it more palatable.

It is possible, even after you witness mating, that the doe will not conceive. There are a variety of reasons: sterility in the doe because of age or a long period of nonproduction, sterility of the male, disease, molting, she is not in heat, or a false pregnancy. False pregnancy (pseudopregnancy) may result when a doe is mounted by another doe prior to breeding or having a buck in an adjoining cage. This stimulates the release of eggs, but obviously no conception will take place. During pseudopregnancy, the doe will go through all of the predictable behavioral patterns associated with pregnancy, but the end result is no kits are produced. This is one good reason not to house does in the same cage.

Because of the above cases of nonconception, it is helpful to be able to tell if a doe is pregnant earlier than the 31st day, so you can adjust her feed and rebreed her as soon as possible. You can feel the young most easily in the doe on the 14th day after breeding. (After that they are too large, and indistinguishable from the internal organs by all but the most expert.) It is best to test her inside the hutch so as not to unduly upset her. Restrain her with one hand, holding the ears and the scruff of the neck as in force mating; with the other hand, gently feel the abdominal area immediately in front of the hind legs. At this stage the young are about the size of marbles, and with some experience you should be able to detect them. To practice, use a doe you know is not pregnant and compare her with one you are sure is. In time, the testing will be routine. A less reliable method is to reintroduce her to the buck 2 weeks after the first breeding. If she growls or whines and otherwise rejects the buck, chances are she is pregnant. But this is not a sure thing—remember, even if a doe is not pregnant she may act the same way if she isn't in heat.

If she appears not to be pregnant, the doe should be bred again. Even if you're sure she isn't pregnant and she accepts the buck again (pregnant does have been known to be mounted even while pregnant), put the nest box in a few days before the original kindling date just to be safe. It won't hurt, and people have had litters from a "dry doe."

Rebreeding

In the wild, rabbits come into heat only in the fall and spring. With domestic rabbits, as mentioned earlier, there is a four-day nonfertile period in each 16-day cycle, but otherwise conception can occur any time during the year. The most normal time to rebreed is after a kindling and 8-week nursing period. You shouldn't wait too long between weaning and mating, as sterility may result. The rebreeding schedule mentioned above would enable a doe to have four litters a year, climate or warmth of your hutches permitting (31 days gestation + 56 days nursing = 87 days; 87 x 4 = 348). If an insulated or heated nest box still doesn't allow you to have successful winter kindling in your part of the country, you will be limited to a maximum of three litters a year: the first to kindle in early April (or earlier if your weather warms up before then), and the third to kindle in late September or early October.

Rebreeding can be carried out as early as the fourth week after kindling (take the doe to the buck's cage as in normal mating), allowing a possible maximum of six litters per year, depending on your doe. Don't try to force her or she won't remain a viable breeder for very long. If she is healthy and chunky, you can rebreed her early; if she is down in condition, wait at least until she's weaned the litter or until she is back in condition. With experience and the daily handling of your rabbits, you'll be able to determine what is best for your herd and your does.

In the case of a lost litter, the doe's feed should be cut to regular maintenance ration and she should be supplied with roughage to help her dry up. She can usually be re-mated three days later. When a doe has only one or two surviving young, they can be given to another doe and she can then be dried up and rebred in a day or so.

Experiments in breeding can be fun and rewarding, if you begin to markedly improve your herd through selective breeding. An acquaintance of ours reported success with breeding purebred medium-weight breeds (New Zealands, Californians) to a giant breed such as Flemish or Checkered to produce larger fryers maturing in the same time as pure medium-weight rabbits, Experiment!

Bucks

Your buck is half your herd, and you should choose him with that in mind. Look for all the qualities you seek in your young stock. In addition, look for one with a shiny coat and bright eyes. Weight is not as critical as it is with does, but a buck that is overly fat may be lazy and not service your does. When that happens, put him on a diet and his interest should return. The scrotum should be full and large and contain two fully descended testicles. If a buck has small and withered testicles (most common in older rabbits), he may be sterile or sire small litters. Bucks of the medium-weight breeds mature in 6 to 8 months, but it is best to wait until 8 months of age before using him. The larger breeds don't mature until at least 10 months. The first few times, the buck may benefit from forced mating, especially when older does are used. As a rule of thumb, a buck can service up to ten does. Ideally, he should not be used more often than every 4 to 7 days.

HEALTH

Rabbits are among the most disease-resistant of livestock, provided they are well-fed and live in clean surroundings. Prevention is once again the byword. Frequent cleaning and disinfecting of feed and water dishes and removal of spoiled hay, frequent cleaning of the hutch, and frequent removal of manure from underneath the pen are all good practices. A solution of bleach is a good disinfectant. Use caution when using a bleach solution; it is extremely caustic and can cause serious burns, especially when it comes in contact with the eyes.

Clean nest boxes after removing them from the hutch and dry them in the sun. Sick rabbits should be isolated for 2 weeks after symptoms disappear. New rabbits entering the flock should be isolated for 30 days to insure that they are not bringing in any disease. The table in the Appendix should be useful in the diagnosing and treatment of the most common rabbit afflictions.

Good ventilation is as critical in rabbit production as it is for other types of livestock grown under conditions of semi-confinement. The air in the rabbitry must be free of ammonia fumes and carbon dioxide. These noxious gases have an extremely adverse reaction on the rabbits' respiratory tract, and their continued presence can lead to severe respiratory conditions. Rabbits need 14 hours of light daily to keep the breeders active and fertile.

The barn or hutch temperature should ideally be maintained at 50 to 75 degrees Fahrenheit year round. By doing this, it should be possible to raise a 4- to 5-pound rabbit on between 1 and 3 pounds of feed. At lower temperatures, too much feed is used to maintain body temperature, while at high temperatures intake is depressed.

In a small operation, achieving this range may be difficult, but effective use of fans and sprinklers should help to reduce the effects of extreme summer heat. During the winter, it is more difficult to achieve the lower end of the recommended range without affecting good ventilation.

If they show signs of excessive heat stress, animals should be moved to an area where the ventilation is excellent. They can be placed on a feed sack moistened with water, and it is also possible to place containers filled with ice in the hutches.

Sprinkling the tops of hutches with water will serve to lower the temperature 6 to 10 degrees on a hot day, when the temperature in a nesting box can become most uncomfortable. Measures should be taken to make the babies more comfortable. Cooling baskets are wire boxes constructed of hardware cloth that can be suspended from the roof of the hutch to allow for maximum air circulation around the youngsters. At night, return the young to their nesting box, but if the outside temperature is still high, they can be returned to the cooling box.

Diseases Caused by Viruses
Infectious myxomatosis: This condition is a fatal viral disease of the domestic rabbit, angoras, Belgian hares, Flemish Giants, and the European wild hare. The poxvirus is transmitted by mosquitoes, biting flies, and direct contact from the California brush

rabbit, which acts as the natural reservoir. The disease is most common on the West Coast of the United States and is seen at the height of the mosquito season. It tends to flare up about once every 8 years, and then becomes somewhat quiet in the intervening years. Animals that develop the acute form of the disease may die within 48 hours.

Animals that survive for longer periods eventually develop ears that become filled with fluid that causes them to droop. There is no available treatment and no vaccine.

Papillomatosis: This disease occurs in two forms in the United States. One form is seen as small, warty growths under the tongue and on the floor of the mouth. The other type is seen as horny warts on the neck, shoulder, ears, and abdomen and is seen mostly in cottontails.

Diseases Caused By Bacteria

Pasteurellosis: Probably the most common bacterial disease of rabbits, the organism that causes it bears the imposing name of *Pasteurella multocida*. This organism can cause a wide variety of different conditions: snuffles, pneumonia, middle ear infections, eye infections, abscesses, genital infections, and septicemia.

Snuffles or nasal catarrh is an infection of the membranes of the air passages and the lungs. It can vary widely in its severity. It usually occurs when the animal's resistance is lowered for some reason. Allowed to progress untreated, it can develop into pneumonia. Animals that recover can act as carriers, accounting for the spread to other animals.

Abscesses arise particularly in bucks that are housed together and fight, giving rise in turn to abscess formation at wound sites, which can be almost anywhere on the body. In many instances you should eliminate infected animals, since abscesses may develop into an infection called septicemia, which usually results in death.

The organism may locate in the genital tract, where it causes varying degrees of inflammation. Animals of either sex that become infected with this organism have a very poor future as breeders.

Mastitis: An infection of the mammary glands, this disease may be caused by a number of possible organisms, but in particular *Staphylococci* and *Streptococci*. Infected glands are hot, swollen, reddened, and painful early in the course of the disease. Later, if left untreated (or sometimes even if treated), the glands may become blue in color, giving rise to the common name "blue breasts."

Treponematosis: Caused by a spirochete called *Treponema cuniculi*, which infects the reproductive tract of both males and females, this condition is identified first by the presence of ulcers and then scabs in and on the external genitalia. Affected animals should not be used as breeders, as the organism is transmitted by mating. It is probably also spread from the mother to her young.

Penicillin is used to treat this disease. Give once a week, subcutaneously. All animals in the operation must be treated, whether or not they are showing visible signs. Once the scabs have healed (in about 2 weeks, on average), the animals can once again be used for breeding. Penicillin may cause enough changes in the flora of the gut to initiate diarrhea. Watch the hutches for this possibility.

Hutch or urine burn: A condition that may closely resemble treponematosis is hutch or urine burn. The only specific way of differentiating the two is by laboratory examination of tissue or material from the affected areas, which are the anus and the external genitalia. This condition occurs when the hutch floors are allowed to become wet and dirty and the animal's rear quarters become irritated. Keeping the floor clean and dry will do much to prevent this condition from developing. Local application of an antibiotic ointment will assist in clearing the infection.

Enterotoxemia: An acute disease characterized by explosive diarrhea, enterotoxemia occurs especially in rabbits 4 to 8 weeks of age. This disease is caused by an organism called *Clostridium spiroforme*, and members of this group are notorious for causing a wide range of intestinal toxemias in a wide variety of animals, especially young ones. There is a relationship between diet and the development of this disease. Where diets high in fiber are fed, far less enterotoxemia

is seen. When conditions in the gut are just right, this group of organisms grow and produce a specific toxin which is highly fatal, even in very small quantities.

Mucoid enteropathy: A disease of unknown cause that can be seen in rabbits of any age, mucoid enteropathy is the result of constipation. Affected rabbits show a variety of signs including gelatinous or mucus-covered feces, being off feed, a dull, poor hair coat, subnormal temperature, and sometimes a bloated stomach due to excess water in that organ. Treatment is not very effective, but sometimes electrolyte solutions are helpful.

Tyzzer's disease: Another disease of young rabbits characterized by a profuse diarrhea and death, this condition is caused by a bacteria called *Bacillus piliformis*, which is eaten where the level of sanitation is poor and the level of stress is high. Very likely a poor level of sanitation can give rise to a high level of stress.

Parasitic Disease

Coccidia: Rabbits can become infected with coccidia, usually at a young age. The organism can set up housekeeping in either the liver or the intestine and is usually swallowed with contaminated feed or water.

The severity of the disease in the liver form seems to be dependent on the number of organisms swallowed. The more organisms taken in, the more severe the disease. It is not generally possible to diagnose this disease until the animal dies or is sacrificed to make a diagnosis. There often are no great outward signs or symptoms that would give the owner a clue as to what might be going on except that the animals just don't seem to do well. Diagnosis is based on post-mortem findings and finding the organism in the bile ducts.

Sulfaquinoxaline is a sulfa drug that can be used either for treatment or as a preventative in the drinking water or in the feed. Check with your veterinarian for appropriate dosage levels after a diagnosis. This drug can typically be administered up to within 10 days of slaughter, but keep yourself appraised as to the current regulations regarding withdrawal times. These regulations are changing rapidly at the time of this writing. As with many disease organisms, coccidia are often resistant to this drug after having been used for many years. One of the drugs suggested as a replacement is toltrezuril, marketed as Baycox.

The intestinal form of the disease is acquired in exactly the same way as the liver form, but diagnosis can be accomplished by having your local veterinarian do a fecal exam. As with the liver form, the symptoms shown by the affected animals are often very vague, typical of "poor doers." Treatment is the same as for the hepatic type.

External Parasites

Ear mite: Probably the most commonly recognized external parasite of rabbits, the ear mite has a dramatic effect on the ear. The inner surface becomes coated with a thick, gummy material which arises because of the constant irritation that the mites

cause, coupled with the severe beating the rabbit inflicts on itself through constant scratching as the result of its misery.

When attempting to clean the infected ear, it is best to heavily sedate or anesthetize the animal as the cleaning is a very painful process. A cleaning with hydrogen peroxide to remove the heavy waxy material, followed by the application of a good ear mite medicine that might be used in cats or dogs is needed. It may be necessary to repeat the ear mite medicine application two or more times to eliminate the mites.

Mites are easily transmitted by direct contact, so segregation of infected animals is essential to control the problem. Thorough cleaning and disinfection of hutches is critical to a good control program.

Non-infectious conditions

Moist dermatitis or wet dewlap: This condition develops when mature rabbits, especially females, get their dewlaps wet from constantly dropping into the water bowl. When this area is allowed to continuously remain moist, it may lead to inflammation, which in turn may become infected. When this happens, it is necessary to clip the fur in the affected area and treat it topically with either an antiseptic or antibiotic dusting powder.

The use of automatic waterers will generally prevent this condition from developing.

Hair chewing and hair balls: These conditions can become a problem if the amount of fur swallowed is excessive. Generally, the rabbit is able to pass the hair through the digestive tract. If this does become a problem, the rabbit tends to go off feed, lose condition, and may die in a few weeks. Unlike some other animals, the use of mineral oil and laxatives is ineffective. Keeping the fiber level of the diet at an appropriately high level (20 percent) will assist in keeping this condition from developing.

Sore hocks: This is a problem especially in the heavy breeds that are raised on wire floors, especially when the wire may have been improperly installed, allowing the soldered joints to protrude upward. Actually, it is not the hock that is involved but rather that part of the rear foot below the hock. Allowing urine-soaked feces to accumulate on the wire adds a further complication. Rabbits with this condition tend to sit with their weight distributed toward their front quarters to take the weight off the sore, inflamed rear feet. When all four feet are involved, affected animals tend to walk on their toes.

Malocclusion: This condition occurs when the front teeth do not come in contact with each other in the proper way and are allowed to grow too long. Usually the teeth in the lower jaw protrude outward and do not come in contact with those in the upper jaw. The teeth of all rabbits continue to grow throughout life, and the rabbits keep their teeth ground down by chewing on wood and other materials. If the problem is getting out of hand, the protruding teeth can be cut back with a pair of small wire-cutting pliers.

FIG. 2.8: POSITIONING OF RABBIT PRIOR TO BREAKING ITS NECK

This condition is considered to be hereditary, so affected animals should not be used for breeders.

BUTCHERING

While some may reason that you should wait until a rabbit matures to 10 pounds before butchering, this is an unsound practice. The most efficient growth occurs during the first 8 to 10 weeks of life; thereafter, feed conversion is not nearly as efficient. While you might get more meat by waiting, it is much more expensive, based on feed consumption.

Once you butcher a rabbit, you may never again consider eating your own chickens. The meat is delicious, better for you, and there is very little waste (20 percent). No more hours of plucking, and feathers up your nose or in your shirt. Once you get the hang of it, you can butcher ten rabbits in the time it would take to do a couple of chickens.

The first time you kill and dress a rabbit will be the most difficult. You'll need a large sharp knife and a plastic garbage bag. Drive a nail part way into a small piece of board, hang the plastic bag from it, and hang the board on a wall. You're ready to begin.

Hold the rabbit by the hind legs in one hand, as shown in Figure 2.8, then grasp the head and quickly and forcefully snap it down while twisting it back. This breaks the rabbit's neck and kills it.

Cut off the rabbit's head, then hang the rabbit by one rear foot from the nail in the board with the body over the bag. Let it bleed out completely.

Cut off the free rear foot and both front feet. Remove the tail and slit the skin (being careful not to cut the flesh) from the hock of the remaining foot down the inside of the leg to the base of the tail. Continue up to the stub of the outer leg. Now pull the skin down off the body, inside-out like a glove. A knife may be used to separate the skin from the fat, which should remain on the carcass.

Slit the belly from the middle of the breastbone to the tail. Remove the entrails.

Save the heart, kidneys, and liver, being sure to carefully remove the gallbladder. Cut off the remaining hind leg, then wash the carcass with cold water to remove any blood and stray fur.

As with most meat, cutting it up is easier if the meat is well chilled, although this is not as necessary as it would be with a larger carcass.

A rabbit carcass, cut up, will have seven pieces. Cut off the front and back legs at the ball joints, then cut the back into three pieces, much like a chicken.

The hide is worth little, commercially. If you want to use it, you can tan it by following the directions outlined in the Appendix.

While 8-week-old rabbits are most often butchered, those older rabbits you cull can also be used. Bucks are generally tougher than does, but they can be stewed or ground and added to pork for sausage.

CHAPTER 3

Sheep

While sheep may not approach the intelligence level of many animals, they can teach them a thing or two about feed efficiency. Like goats and cows, sheep are ruminants, or animals with a four-compartment stomach which, through microbial action, can convert cheap roughages and other inexpensive low-protein feeds into high-quality nutrients that are high in protein and other essential components of a well-balanced diet.

Sheep are hardy, surefooted animals that can graze the hilliest, rockiest pastures unsuitable for other livestock. Because they eat a wide variety of roughages, they are excellent weed killers and this, along with their rich manure, makes them an ideal animal to improve poor pasture. (They also make superior lawnmowers—much better, I think, than those models that need gasoline.)

Lambs purchased in the spring will fatten quite nicely to butchering weight by the fall, on grass alone. They need relatively little grazing space, and by using a tethering system and being careful, a few sheep can do well without any fencing.

Sheep will also supply you with another valuable product besides meat: 3 to 18 pounds of wool per year that can be used for spinning, knitting, or sale, to help defray the costs of your sheep and other livestock. Purchase a few ewes and a good ram and you will be supplied with lambs year after year, their cost being only a fraction of the cost of your ewe. By starting your own mini flock, delicious lamb and useful wool will be yours for a lot less than the consumer pays.

Table 3.1: Classes and Common Breeds of Sheep					
FINE-WOOL TYPE	**MEDIUM-WOOL TYPE**	**LONG-WOOL TYPE**	**CROSSBRED-WOOL TYPE**	**CARPET-WOOL TYPE**	**FUR TYPE**
American	Cheviot	Cotswold	Columbia	Black-Faced	Karakul
Merino	Dorset	Leicester	Corriedale	Highland	
Debouillet	Hampshire	Lincoln	Panama		
Delaine Merino	Montadale	Romney	Targhee		
Rambouillet	Oxford	Romeldale			
	Shropshire				
	Southdown				
	Suffolk				
	Tunis				

BREEDS

There are many breeds of sheep, classified into six groups as listed in Table 3.1.

We are most interested in the medium-wool types. They are primarily raised for their meat, being rather blocky in conformation and relatively fast-maturing animals. They are not usually known for their wool production, but recently more emphasis has been placed on this in their breeding, and they are good dual-purpose sheep. If you are most interested in wool, you should look into fine-wool breeds, which are also becoming better meat producers as the result of careful selective breeding. Long-wool types are generally too slow-maturing to be important here, but crossbreeding them with the fine-wool breeds has made them a fair dual-purpose animal, and for that reason they are often classed with the medium-wool types.

If you are planning to purchase grass lambs (spring-born lambs that will be raised on pasture and butchered in the fall), the specific breed will be of little importance to you. The key considerations will be availability in your locale, and personal appeal, although it is important to note that some breeds are faster maturing than others and will lend themselves more readily to a spring-to-fall rearing program. If you plan to do your own breeding, a deeper knowledge of breeds will be necessary. In any event, some familiarity with the common breeds is called for.

FIG. 3.1: EXAMPLES OF SHEEP BREEDS

Medium-Wool or Mutton Breeds of Sheep

Cheviot: White faces and legs covered with short white hair. Black noses, lips, and feet, with short, erect ears. Cheviots are quite hardy, prolific (125 percent lamb crop*), and they make good mothers. They are one of the smaller breeds and mature much more slowly, so are not ideal for a summer-raising routine.

Light fleece (6–8 pounds). Mature rams average 175 pounds; ewes, 125.

Dorset: White-faced, horned, or polled (naturally hornless). Can be bred at any time of the year. Ewes are very prolific (150 percent lamb crop) and are good milkers. Fleece averages 8–19 pounds. Rams average 200–225 pounds; ewes, 150–175.

Hampshire: Black faces, ears, and lower legs. The ewes are prolific and good mothers. This is a large, fast-maturing sheep, and for that reason is excel-

---

* Lamb crops are commonly measured in percent as an average of lambs weaned. 100 percent would indicate one lamb per year weaned; 200 percent would mean twins weaned each year. 150 percent is the mark most breeders shoot for.

lent for summer raising, especially in cooler regions where the pasturing season is short. Fleece averages 7–8 pounds. Rams average 250 pounds; ewes, 180.

Montadale: Small heads, open, white faces, clean legs. Montadales are prolific and good mothers. They are a very good dual-purpose breed with heavy (10–12 pounds) fleeces. Rams average 250 pounds; ewes, 175.

Oxford: A large sheep with gray to brown faces, ears, and legs. A fast-maturing breed whose ewes are prolific and good milkers. Rams average 300 pounds; ewes, 175–250. Heavy (10–12 pounds) fleece.

Shropshire: Deep brown or black feet, noses, ears, and legs. Wool covering faces often results in wool blindness, which is a drawback, but breeding is aimed at eliminating this. A good combination wool/meat breed that produces a large carcass and a fleece that averages 9 pounds. Rams weigh 175–200 pounds; ewes, 135–150.

Southdown: The most desirable meat breed. Blocky and short-legged with gray to brown faces. Southdowns are small but mature very early. The fleece is light (5–7 pounds). Rams weigh 175 pounds; ewes, 125.

Suffolk: Open, black faces; black legs, and ears. The ewes are good milkers and very prolific (150 percent lamb crop) and have a fleece weighing between 8 and 10 pounds. Rams average 250 pounds; ewes, 180.

Tunis: This is a hornless (polled), open-faced sheep, with a tan or red face. The breed is fairly rare in this country A good milker. Lambs are early-maturing. Rams weigh 150 pounds; ewes, 110–125.

It is not really desirable to purchase purebred sheep, except perhaps in the case of a ram. Purebreds cost more, will not grow faster than good-grade sheep, and may grow more slowly than crossbred animals. As with other animals, they may lack the hybrid vigor associated with crossbreeds.

PURCHASE

In choosing a breed to start your flock, you should be influenced by the following factors: personal preference, availability, products desired (wool or meat), and, to a lesser degree, weather and other environmental conditions.

You will do best to choose a breed that is pleasing to you over other factors, since it will not be easy to sustain interest in a breed that is personally unattractive. If you want both wool and meat, one of the medium-wool breeds should fill your needs. Interest in sheep for wool only and not slaughter would lead one to choose from a breed where wool production is higher. If you have sparse or rocky and hilly pasture, you should consider one of the hardier breeds of sheep. In areas with severe and snowy winters and where undergrowth is thick, closed-faced breeds will be more susceptible to wool blindness.

In buying your sheep, avoid livestock auctions unless you are truly an expert at selection, which indeed most of us are not. Your best source is a reliable sheep farmer. A sheep farmer will be better able to understand what you are looking for and should have a greater number of animals to choose from and also give you some helpful start-up tips.

Here are some of the things you will be looking for: grass lambs to raise over the summer for butchering in the fall, ewes for breeding, and a ram or rams for breeding.

Grass Lambs

Buy your grass lambs as soon as you can after the grass reaches grazing length in the spring. This will give you a longer pasturing season and more lamb for your money. You will be offered ram lambs or wethers (castrated males). Ewe lambs of good breeding quality generally are not sold for meat but are reserved for future breeding and are generally higher in price. While a wether will weigh slightly more than a comparable ram after a few years' time, for the purposes of raising over the summer, these differences in carcass and taste are negligible. If you already have breeding stock, however, purchase wethers or castrated spring lambs, or you will have considerable problems trying to keep your potential lamb chops from breeding your ewes.

Be sure that your lamb's tail is docked; this ensures cleanliness and helps prevent infection or infestation with maggots.

Look for: Large, well-formed lambs that are straight-backed and stand squarely; a deep chest; good width between front legs, which are well muscled and firm; alertness; and lambs that walk with ease. Part the fleece, and look for bright pink skin, a sign of good health. The mucous membranes around the nose and mouth should also be a healthy reddish color.

Avoid: Generally the opposite of above: small, listless, or generally unhealthy looking lambs; or those with diarrhea, colds, or runny noses. Avoid poorly proportioned lambs and those with dark, bluish skin, or those with crooked legs or which demonstrate lameness. A sheep whose front legs are so close together that they appear to emerge from the same cavity may indicate severe inbreeding. Do not get into the trap of falling for the runt, with the notion the you are going to make it into something it does not have the potential to become. This warning applies to any animal that you may be considering for purchase.

Note: If you add grass lambs to your current flock of older breeding ewes, possibly including a ram as well, savage bullying of the lambs could result. We have discovered that adding grass lambs in pairs gives the newcomers some respite and helps alleviate this problem.

Breeding Stock

Ewes: In purchasing your breeding stock, considerably more care is called for. While a grass lamb is raised for perhaps 6 months and then slaughtered, a ewe used for breeding will hopefully be around for several years, and as a result will affect your entire flock. You can buy a ewe lamb in the spring to raise for breeding purposes, or purchase older ones in the late summer or early fall. Especially for the beginner, it is advisable to buy a ewe at 2 or 3 years of age, that has already lambed. Such an ewe will be mature and have had lambing experience, which will mean fewer demands on the novice owner. If possible, buy one that is already bred so you can start off the first year without having to purchase a ram and worry about arranging a successful breeding.

Avoid livestock auctions and private owners who may try to foist their sickly ewes or poor lambers on you. A sickly ewe can barely support herself, much less carry a lamb and have enough milk to nurse it to weaning age. We (the Thomases) have had good bargains buying culls from larger operations. These are not 'culls' in the usual negative sense, but normal, healthy sheep from a very large flock that do not compete well for feed in larger groups. In our small flock, they thrive. What is a loss for the owner of a large flock is your gain, at a bargain price.

Look for the qualities you want in a good grass lamb. Look also for relatively large ewes, because large animals tend to have larger, faster-gaining lambs. Twinning is an inherited trait in sheep; therefore a ewe that is a twin, or has produced twins, is more likely to have offspring that continue to carry that desirable trait. Avoid shallow-bodied animals or ones that are long-legged or those with narrow heads. Again, these points are relative, and experience will help you to evaluate your stock. The fleece should be uniform, compact, bright, and clean; avoid animals with uneven coloring. The sheep should walk easily, and the hooves should be well formed and show no evidence of cracks in the sidewalls. Foot problems will affect how well a ewe can carry her lamb. Avoid those with foot rot, as well as those that are limping.

In large commercial flocks, sheep of the medium-wool breeds don't begin to decline physically until they are about 6 years of age. Older sheep cannot feed well when grouped with younger, stronger ones. As explained earlier, older sheep that are culls from larger flocks can be a good addition to your small flock, if the price is right. With the proper care given by a conscientious stock owner in a small flock, they can usually lamb for a few years after they have "washed up" in the larger flocks.

Avoid young ewes (2 to 3 years of age) that haven't lambed yet. If a ewe hasn't lambed by 3 years of age, she very likely has an underlying reproductive problem. With practice, you will learn to spot such animals. If the ewe's teats are very small (relative to her age), it is probable that she hasn't had a lamb—or at least one that's lived long enough so that she could nurse it. If there is any doubt in your mind, don't buy.

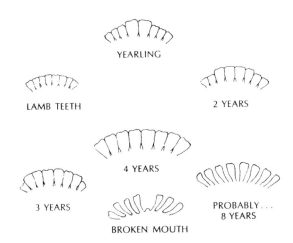

FIG. 3.2: AGE DETERMINATION BY STAGE OF TEETH DEVELOPMENT *(See text)*. (From *The Production and Marketing of Sheep in New England*, a New England Cooperative Extension Publication)

With all this discussion of age, in considering buying stock, it would be helpful to know how to determine it. If the sheep is a purebred with papers the task is quite simple. If there are no papers, the development of the teeth is a fairly accurate, albeit not infallible, indicator of age (Fig. 3.2). A lamb has narrow teeth, corresponding to the baby teeth in humans. At the age of 12–18 months, the centermost teeth are dropped and replaced by two broad permanent teeth. Each year thereafter, an additional pair of permanent teeth appears until, at age 4, the animal has what is known as a "full mouth."

After this, the determination of age by teeth is less accurate. Often, the full set of teeth is intact until the age of 8 or 9, but more often a few teeth are lost around age 5 or 6, resulting in a "broken mouth." Older sheep with no teeth are known as "gummers" and (no surprise!) are not in high demand. With experience, the approximate age of older sheep can be determined by the degree of wear or spread of the teeth, and the number of teeth lost.

In a sheep with a full or broken mouth, fitness and age can further be determined by the condition and placement of the rear teeth. By feeling toward the rear of the mouth (Caution: sheep have strong jaws; you should always have another person hold the mouth open while you do this), one can check the condition and shape of the rear teeth. Those with teeth that are still intact and meet well are apt to be of producing age, while those with missing teeth and abscesses should be avoided.

Perhaps the most important single consideration in the purchase of a ewe is the quality of her udder and teats. A ewe, no matter how many twins she drops, won't be worth much unless she can provide them with an adequate supply of milk. The udder should be soft, pliable, and free from lumps. Avoid animals that have abscesses or ruptures of the udder and blind or missing teats, or those whose teats are abnormally large or thick.

Rams: In choosing a ram, spend even more time, often beginning to look many months in advance so that you'll find a suitable one and not get stuck with a poor substitute. For, while a ewe passes her traits on to one or two lambs per year, your ram will affect every lamb in your flock. In ewes, you should get good grades or

crossbreeds, but for a ram, spend a few extra dollars and get a purebred. In this way you'll be constantly upgrading your flock and imparting hybrid vigor. At worst, use a grade or crossbred ram that embodies as many desirable qualities as you can find (rate of gain, size, twinning, etc.), but remember that you won't be upgrading the quality of your sheep. If you breed with a ram that is better than your flock, in the long run you will get a better flock. By breeding to a ram that is equal to or poorer in quality than your other animals, you will downgrade your flock or at best not improve it.

In addition to the desirable qualities mentioned earlier in this section, select a ram that is large and thick, with plenty of bone. Be aware of inherited traits listed in Table 3.2, which is also applicable to ewes. Choose a ram that is "full of it"; that is, active and rambunctious. He should be between 1 and 6

years of age, with strong legs, and wide at the ears. Also, last, but most certainly not least, make sure your ram has two fully developed testicles.

HOUSING

Sheep are perhaps the most defenseless of farm animals, and fencing and housing are as much to keep predators out as to keep the sheep in and protect them from the weather. If you plan to raise only grass lambs, and not winter or breed them, and your area is relatively free from stray dogs and other predators, you're in luck. You can get by with just tethering your sheep, being sure to provide them with shade in warmer weather.

In many parts of the country today, wildlife in a variety of different forms is making a comeback. The suburbs are becoming a favorite haunt of coyotes, fisher cats, and other predators that can make sheep raising a high-risk enterprise.

Tethering

In tethering, you attach a chain to a collar on the lamb, and then to a post with a swivel on it. With a swivel, the lamb can move about with a minimal chance of wrapping the chain around the post. I have seen large anchors intended for dogs that screw in flush with the earth. These will serve the same function as a stake, but eliminate the biggest danger in tethering: the lamb wrapping around the post and choking itself.

Steve Thomas lived on a road with little traffic and no stray dogs, and in the first year had two sheep and was able to leave them loose most of the

Table 3.2: Heritable Characteristics in Sheep

MEAT CHARACTERISTICS	% HERITABLE	COMMENTS
Twinning	15	Twins increase pounds of lamb sold.
Birth weight	33	Larger lambs at birth usually make faster gains.
Weaning weight	33	Weaning weight is tied to cost of production per head.
Yearling weight	43	This is the ability to gain on own, without dam.
Post-weaning daily gain	7	Gain without mother's milk is efficiency of animal to produce on own.
Efficiency of gains	15	Pounds of feed per pound of gain important in economy.
Body type	12	Not highly heritable, but affects purebred value and market value.
Condition score	12	Important in lambs sold at weaning.
WOOL CHARACTERISTICS		
Face covering	43	Ewes with open faces produce 11.1 more pounds of lamb per ewe than closed-faced ewes.
Neck folds	35	A defect in certain breeds.
Grease fleece	47	On average, 33% of income from sheep is from weight of shorn wool.
Staple length	45	Major factor in determining fleece weight.
Fiber diameter	57	Important in determining wool grade.

Prepared by Byron E. Colby, Veterinary and Animal Sciences Department, University of Massachusetts; quoted in part from *Genetics of Livestock Management* by John F. Lasley and *The Stockman's Handbook* by M. E. Ensminger; and made available by the Vermont Extension Service in cooperation with the Vermont Sheepherders Association, 1975.

time. A handful of grain kept them hanging around every morning but they never wandered far, and put themselves to bed in the barn every night. This "honor system" worked well until a friend's child chased the lambs one day, and they suddenly realized that they could go anywhere they wanted. From then on, tethering was a must. With most sheep, you can tether one and the flocking instinct will keep the others in sight. With a larger number of sheep, the mathematics become more complex. One summer with a flock of five sheep, we could tether three and keep the other two around, but if we tethered only

two, the other three would wander off. Apparently sheep can add and subtract.

As we are discussing sheep wandering off, it is worthwhile to recommend selecting one lamb—perhaps the most rambunctious, and thus the most likely to escape—as the "bell lamb." This then is the derivation of the term "bellwether." (To further show the influence of the sheep in our history, consider the manner in which some towns were named, as in Wethersfield, Connecticut.) If they wander, you can hear the bell at some distance. Try finding a lamb without any bell in the woods, and you'll know what I mean.

If you tether sheep you can't just leave your lambs unattended on the "back forty" until butchering time in the fall. It works best when they are close to the house and you can see them easily—one of the pleasures of sheep. They must have shaded spots to retreat to during hot weather. Be sure there are no obstacles such as trees, heavy weeds, and briers, in which they may become tangled. If you tie them out during the day and lock them up at night as a precaution against predators, realize that during hot weather they often eat nothing during the day, choosing to graze at night when it is cooler. If you enclose them at night, they will gain slowly or even lose weight. You must always be on guard against stray dogs making quick work of a defenseless lamb, rendered even more defenseless by tethering. If dogs are a problem, you'll probably have to go on to more elaborate and expensive methods of containment.

Check the lamb's neck periodically for any sores or chafing from the collar. If not treated, they may become infected and infested with maggots. Most of all, check your lambs daily and unwind their chains so they do not become wrapped around the stake. Tethering is a very good and cheap system for a few sheep *provided that you are vigilant.*

Woven Wire Fencing

Sheep fencing is one of the more negative things about keeping sheep, particularly because it is expensive. I wouldn't even consider fencing large areas unless I were going to breed and winter sheep. We've tethered up to five sheep at a time, and that was our limit. It was more than we could safely handle—in fact one hung itself on a tether that summer. At the least, you might buy one small roll of woven wire fencing to confine and protect your sheep at night or when you are away. Another option is to move that roll of fencing around as the sheep graze down one area. This is fairly simple and will provide confinement and protection relatively cheaply.

For permanent fencing using woven wire, we have what I term "conventional" and "deluxe" systems. The conventional fence, and one that will do nicely in most instances, is 39- to 48-inch woven wire attached tightly to stakes so dogs can't get under it or over it. If you still have problems with dogs or your sheep get out, the deluxe system, patterned after the USDA design, has 32-inch high 6-inch-stay vertical woven wire fence with one strand of barbed wire running along the bottom and two to five strands at the top, giving the fence a total height of 5 feet. This will keep out all but the most aggressive dog and most people, and will cost accordingly. If you really want to go into sheep farming, this is the fence for you. But most people will get by with the conventional setup. Young lambs can sometimes slip through the holes in this fencing and may wander off. To prevent this, make them wear a triangular yoke made out of lath. It should measure 8 to 12 inches on a side. This will prevent them from slipping through the fencing and will not hamper nursing.

Electric Fencing

Most people give sheep very little credit for intelligence, perhaps deservedly, but they can be taught to stay behind an electric fence. This can mean big savings for those who want to fence a large piece of land. It is considerably cheaper than woven wire and easier to install. However, it does require frequent maintenance and checking, and if you live in a snowy region, will require resetting every spring.

For most purposes, a two-strand fence of 12- or 14-gauge galvanized wire set at 12 to 14 inches and 22 inches will be adequate. Young lambs may still get under such a fence, so if their wandering becomes a problem you'll have to go to a 3-strand fence set at 8 to 10 inches for the first wire, 26 inches for the top, and split the difference for the middle wire. An electric fence will generally keep dogs out and sheep in. The key factor is that the wire must be *very tight* for sheep (more so than for any other animal) or the wire will give, and not push through the fleece to the skin. The fence charger should be the finest available and properly installed. As with any electric fence, it is necessary to check it periodically to make sure nothing is grounding it and causing it to lose its charge.

Sheep must be trained before putting them behind this type of fence to make them respect it and to prevent them from trying to get through it the first time out, thus getting entangled and shocked and subsequently breaking it in their panic. Train them in the spring after shearing, and ideally after a heavy rain has soaked both the sheep and the ground. Make an enclosure about 3 to 5 feet by 5 to 10 feet, fenced with electric wire, and set the charger on it. Place a pan of grain in it and let the sheep in and allow them to attempt to get the grain. Watch them closely and be quick to repair the fence if it breaks down or is grounded. Young sheep will learn more quickly than older ones who have never felt an electric fence. This, and occasional jolts from their permanent fencing, should render them fence-wary until the next spring. Sheep with a full coat can go through an electric fence just like a runaway train and, if this is allowed to happen, training them to respect the fence becomes ever so much more difficult.

Shelter

If you are going to raise only grass lambs you will have to provide shade for them, but no building or enclosure will be necessary unless predators are a problem. For wintering sheep, especially in colder climates, some sort of enclosure is required. It can be a three-sided lean-to, open to the south or southeast; or, if early lambing is desired, it should be enclosed and relatively draft-free. For later lambing, or in warmer climates, lambing can take place outdoors and a simple lean-to will suffice. Do not cramp your sheep or force them to live in a filthy, foul-smelling pen. Allow at least 16 square feet of floor space per ewe, and furnish good bedding (sawdust, wood chips, straw), and keep the pen scrupulously clean. Clean pens not only make for healthier sheep, but also make for better-tasting meat. The reason much commercial lamb has that heavy, "lamby" taste is that the animals are often forced to live in cramped quarters and sleep in their own excrement, which taints the meat.

Door openings should be at least 8 feet wide to prevent sheep from becoming crowded in doorways and injuring themselves and any unborn lambs. It is important to have electricity in the pen for a heat lamp, when lambing early. Lighting will also make working in the pen, especially during lambing season, much easier. Have an outside yard off the sheep barn so that in winter the sheep—especially the ewes—can exercise.

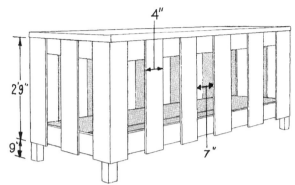

FIG. 3.3: A COMBINATION HAY AND GRAIN FEEDER.
A hinged top provides protection from the weather.

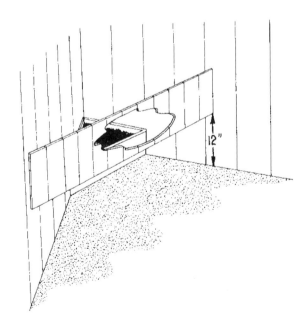

FIG. 3.4: A SIMPLE CREEP FEEDER FOR LAMBS

We have our barn enclosed within the fencing and feed our sheep outside in the barnyard all winter. In this way, they get plenty of fresh air and exercise.

Lambing Pen

If your ewes lamb when the sheep are back on pasture, no special equipment is necessary. However, if lambing is to take place inside, you should have one or more lambing pens. This can be a portable enclosure that you use to confine the mother and her lamb(s) from the rest of the flock for the first few days. This permits them to bond and reduces the chances of a mother losing or abandoning her offspring. The pen should be about 5 feet square, and must be in a draft-free location. If your barn is large you can construct permanent lambing pens. We use ours in the fall to keep our ram (or rams) away from the ewes until we are ready to breed the flock. These pens can also be used to isolate sick sheep from the rest of the flock, to house a veal calf or some weanling pigs—or to confine a neighborhood child who will not quit chasing your sheep.

EQUIPMENT

If you don't have a stream or other running water, you'll have to supply your sheep with a water bucket or a trough. Sheep don't drink as much water as most other livestock will, but a pail of water, kept out of the sun, should be available at all times. When sheep are tethered, they will wind the chain around the water bucket and overturn it. If you're raising grass lambs, all you'll need is a water bucket and a small dish or discarded pan for feeding grain. For feeding hay and grain in the winter, it is best to have a combination hay and grain feeder similar to the one pictured in Fig. 3.3.

The first winter we kept sheep, we fed them hay on the ground, and I would guess conservatively that a third of the hay was wasted: urinated on or trampled underfoot. Although they do not quite equal goats in their refusal to eat off the ground, sheep will waste a lot when fed in that way. Since we began using a manger, there has been almost no wasted hay. Whatever form of manger you use, remember the space between the slats that the head goes through should be 7 inches wide so that the head can fit in but the animals cannot pull out too much hay at any one time. The total width of the manger is variable, but be sure to allow 15 inches for each mature sheep and 12 inches per lamb.

A box or dish, prevented from tipping and protected from the weather, should be available to supply mineral salt to the sheep. This should be supplied in loose form, since salt blocks may chip teeth and subsequently hamper the ingestion of feed. (At times, salt blocks are also acceptable: see Feed section.)

To insure the maximum rate of gain, it is often recommended that you feed your lambs extra grain. You will need a creep feeder—a feeder or a space that is accessible only to the younger animals (Fig. 3.4). The easiest method is to use an existing lambing pen or build one in the corner of your barn, constructing the opening so that it is only 12 inches from the ground. The lambs can get at the grain and hay inside, while the ewes cannot.

FEED

Sheep are desirable because they can grow and produce meat, wool, and lambs by deriving 90 to 100 percent of their nutritional needs from pasture and roughages. In common with the other ruminants (cows, goats, and camels), they can take low-protein feeds and convert them into high-quality food. In the first of the four compartments making up the stomach, the rumen, the feed is processed by billions of bacteria and protozoa, which convert high-fiber diets into essential amino acids, the building blocks of protein. Also, many essential vitamins, particularly the B-complex group, are readily synthesized, thus they need not be added to the ration.

The rumen is nonfunctional in a young animal, so for the first few weeks of life a complete diet, consisting of ewe's milk together with a good quality creep feed, must be supplied. After a few weeks the rumen becomes functional, and the lamb is on the path to becoming a true ruminant.

Grass Lambs

Rams lambs intended to be butchered in the fall are among the easiest of all animals to manage. Offer them good-to-excellent pasture, plenty of water, and a salt lick and let 'em go. Sounds almost too good to be true, and probably is unless you happen to dwell at a latitude and longitude ideally located to provide the sort of growing conditions that will maximize the growth of grass. Unfortunately most of us don't live there, so we have to make some additions to the feeding program in order to ensure that in the fall the lambs we are raising will hang the sort of carcass that will provide us, our friends and neighbors, and close relations with delicious lamb; the sort that mother used to bake for Easter Sunday. In the northeast and many other sections of the country, grass follows a definite pattern of growth and dormancy, a pattern much influenced by rainfall and temperature. In many regions of the United States, grass reaches its maximum nutritive value in mid-to-late May and declines rapidly in value, so that by July its value is rather minimal, with lots of fiber but far less in the way of protein and other essential nutrients. If you are planning to rely on grass alone to feed your lambs over the summer, have some grain and good-quality hay available to fill the gaps that will likely arise along the way. During the heat of mid-July and well into the latter part of August, most native grasses enter a semi-dormant growth pattern during which very little new growth takes place. If you are blessed with a large amount of acreage and have a small number of animals, then you probably will get by, but many of us are not so blessed. We are more likely to have more animals than we should on a limited amount of acreage.

In one respect, sheep are their own worst enemy when it comes to good pasture management. Being close grazers, they tend to nibble the grass to the point where the crowns of the plants are completely exposed to the full power of the sun, thus drying them to the point where early dormancy is likely to occur.

Pasture: A good pasture can handle up to a dozen ewes and their lambs per acre. This is a maximum; most pastures will furnish far less grazing than that. Watch your sheep carefully and note the condition and length of the grass. If you don't have additional pasture, you may have to give up a few sheep (or tether them on your front lawn) until you achieve a balance.

The best system involves at least two pastures, so that when one gets eaten too low you can shift to the other and allow the first to re-grow. This rotation of pastures is also important in the control of parasites. Sheep will improve a pasture in the long run, weeding, fertilizing, and spiking in the fertilizer with their cloven hooves. A pasture that handles only four sheep per acre may handle six in a couple of years. It is also important to keep your pasture in shape with lime and additional fertilizing when needed. Soil testing is the key. Your local cooperative extension office will provide you with all of the information you need, plus details regarding the obtaining and submission of soil samples. The best sheep pastures include alfalfa-brome grass, clovers, clover-and-grass mixes, ladino clover mixed with other legumes and grasses, and orchard grasses. Another good source of grazing is hayland, after the last cutting is harvested in the fall

Watch out for plants that are poisonous to sheep, such as lupine and chokecherry. Usually these plants will not be eaten unless pasture is very low and there is no other forage available, but *Veratrum* (false hellebore) is relished by sheep and can cause birth defects if eaten later in pregnancy. Burdocks can become hopelessly entangled in wool, making it worthless.

Water and salt must always be available to sheep. While loose salt is preferred over salt blocks because of less danger of injury to teeth, I have found that some of our sheep dislike the consistency of loose salt and prefer the salt lick. It is better to give it in that form rather than have them not get any. Two of the most important minerals for ewes are calcium and phosphorus, essential for the growing lamb and the production of milk. Often these minerals are missing from pasture and hays, especially if poor quality hay is used, and must be provided as a supplement. Always supply a mixture of either ½ mineral salt and ½ dicalcium phosphate or ⅓ mineral salt and ⅔ steamed bone meal, given free-choice in weatherproof feeders.

Important vitamin requirements for sheep are A, B-complex, D, and E. If you are feeding good hay with at least N legume in the mix, your vitamin A needs should be satisfactorily met. Commercial grains usually contain vitamin supplements, so check the bag's label if you fear there are deficiencies. Other sources of vitamin A are cabbages, corn, squash, and carrots, which are all relished by sheep. As long as your sheep receive at least a few hours of sunlight a day, their need for vitamin D should be met. Vitamin E deficiency may evidence itself by stiff-legged lambs. If this is a problem, wheat germ oils should be fed, and a veterinarian should inject your lambs with vitamin E. B-complex deficiencies should not be a problem because, as noted earlier, they are synthesized in the rumen of sheep.

Hay: Hays of the types listed for pasturing will be suitable. If a high-quality legume hay or legume-grass mixture is available, it should be possible to save substantially on feed bills by reducing the amount

Table 3.3: Suggested Daily Rations for Ewes in Good Health from Breeding until Six Weeks Prior to Lambing		
#1	Grazing on good pastures may be continued, but care should be taken not to overgraze legume seedlings or they will be winter-killed.	
#2	Barn feed legume hay	3–4 lb.
#3	Grass hay 32–36% protein supplement or Commercial mixed grain (12–16% protein)	3 lb. ½ lb. 1 lb.
#4	Mixed legume and grass hay Protein supplement or Mixed grain	2¼–3¼ lb. ¼ lb. ½ lb.
#5	Legume or grass-legume hay— good quality Corn silage	1½–2½ lb. 4–6 lb.
#6	Corn or non-legume silage High-protein supplement	8–10 lb. ¼–½ lb.
#7	Good grass or legume silage Hay	8–10 lb. 1 lb.
#8	Late-cut timothy hay Corn Oats Bran	3 lb. ½ lb. ½ lb. ¼ lb.
#9	Grass hay Mixed grain (14–16%)	3–4 lb. 1 lb.

Table 3.4: Suggested Daily Rations for Ewes 4-6 Weeks before and up to Lambing		
#1	Legume hay Oats, bran, corn or barley (or mixture of all in equal parts by weight)	3–4 lb. ¾ lb.
#2	Grass-legume hay Corn silage Mixed grain or bran and oats High protein supplement	½ lb. 4–6 lb. ¾ lb. ¼ lb.
#3	Timothy hay High-protein supplement Mixture of corn, oats, and bran or commercially mixed grain	3 lb. ½ lb. ¾ lb.
#4	Mixed legume hay Mixture of oats and bran or bran and corn or commercially mixed feed	3–4 lb. 1 lb.
#5	Corn or non-legume silage Good non-legume hay Protein supplement Mixture of bran, corn, oats or oats and bran, or commercial mixed feed	6–8 lb. 1 lb ¼–½ lb. ¾ lb.
#6	Legume hay Timothy hay Mixture of bran and oats, or bran and corn	2 lb. 2 lb. 1 lb.

Source: Tables 3.3 and 3.4 are from *The Production and Marketing of Sheep in New England a New England*, a New England Cooperative Extension publication.
These rations may be too low in calcium. In this case, or anytime non-leguminous roughages are fed, feed a mineral supplement as suggested earlier.

of grain that needs to be fed (see Tables 3.3 and 3.4). Second- or third-cutting hay, while usually a bit more expensive, is well worth it because sheep relish this finer hay. An axiom to remember is that a good forage program should be the basis for any livestock feeding program Because fewer hard stems are present, less is wasted. Do not force sheep to eat the coarser hay by leaving it in the manger until they clean it all up. It is less nutritious and may be removed for bedding, or fed to pigs or other less choosy stock. A mature sheep will consume 500 to 700 pounds of hay in a winter.

Do not feed sheep a feed concentrate made up solely of corn. Sheep should also not be put out onto very rich clover or alfalfa pasture that is wet with rain or dew, as bloat may result. In the spring, when legume pasture is close to reaching its maximum growth, it is a good idea to give the flock some hay before turning them out onto lush pasture, as bloat may result from the ingestion of too much highly fermentable clover. Having the rumen already partially full reduces the volume of pasture plants that may be consumed, thus lessening the possibility of bloat. As with all stock, feed on a regular schedule. During the time in the spring when leguminous plants are at their optimal growth, limiting the number of hours the flock is allowed to graze is an excellent management tool.

Ewes
Proper feeding of ewes will lessen the probability of stillbirths, and result in healthier lambs that will gain faster. Furthermore, it will prolong the productive lives of the ewes, increase their milk flow, and

lessen the chances that they will reject their lambs because of compromised condition.

As with other stock, flushing (improving the ewes' diet) will improve their breeding efficiency. Flushing 2 weeks before and up until breeding will tend to bring all the ewes in heat at about the same time, hence giving you more uniform lambing times. It has been shown that flushing will increase your lamb crop by about 18 percent. Ideally, you will have a lean ewe gaining weight at the time of breeding. You do not want your ewes to be overly conditioned, as that lessens the likelihood of good-to-excellent breeding efficiency. If your ewes are overweight, switch them to sparser pasture 6 weeks prior to flushing so that when they are ready to be flushed, they will be down to normal weight. Flushing can be accomplished by switching the ewes from fair-to-excellent pasture, or if on good pasture or already on hay, adding ½ to 1 pound of grain to each ewe's feed per day. Upon successful breeding, feed as indicated in Tables 3.3 and 3.4.

It is just as important not to overfeed your ewes as it is to make sure they aren't underfed. Above all, ewes need plenty of exercise during pregnancy. Feeding them outside will force them to exercise and also expose them to sunshine, especially during the winter months when the available hours of sunshine are limited.

Sheep require a grain ration that contains 14 to 16 percent protein. The next most important factor is its palatability to the flock. The generic names for these feeds may be sheep feed, goat feed, or horse feed. These are usually a balanced ration, containing vitamins, minerals, and other necessary foodstuffs. We have found a 14 to 16 percent dairy ration to be both a relatively cheap feed and a very palatable one for our sheep. Experiment with different feeds and find the one most favored by your sheep, and your pocketbook. Good high-protein supplements include soybean, cottonseed, and linseed oil meals. Care must be taken to avoid feeds designed for other species that may contain high levels of copper—levels that are appropriate for that species but not for sheep. Read the labels on feed bags carefully, and if there is any question consult with the feed store manager, or your veterinarian.

Use what feeds you have to make up the most economical mix. Once you choose a program, try to stick to it—if you must change, do so very gradually. When feeding silage, be certain that the quality is good, free of mold or other evidence of spoilage, with an odor consistent with good quality. Sheep do not tolerate poor-quality silage well, and they are prone to a whole range of digestive problems from ingesting such feedstuffs, including death. When instructing veterinary students about the challenges of sheep practice, one acronym prevails: S.S.S.S., "sick sheep seldom survive." Sheep are *not* remarkably good patients.

Up to 4 weeks before lambing, you can save substantially on feed if you have your own legume hay available; the less nutritious your hay, the more you will have to pay for extra feeds. You can also save by not overfeeding your sheep. Up until a month before lambing, the sheep should be well covered over the backbone and ribs, but you should still be able to feel bone in these parts. If you can't feel the bone, you're overfeeding. As Table 3.4 indicates, you should increase feed a month or six weeks before lambing because 70 percent of the growth of the fetus takes place during this time. Follow recommendations. It is just as bad to overfeed during this time as any other. If overfed, the ewe will be too fat, the fetus will weigh too much, and you may have lambing difficulties and possibly a resulting loss of lambs and even ewes.

When signs of lambing are imminent, cut back on bulky feed and add ½ pound bran to the ration as a laxative so there will be less competition for room between the end of the large intestine and the vagina at lambing time. This is especially important for first-year lambers and ewes that have a history of difficult lambing.

After lambing, don't grain for 24 hours, since udder problems may occur and the lamb may scour. Then gradually build the grain back up to 2 to 3 pounds per day, depending on how the ewe milks and the lamb reacts. In the case of twins, which are quite a strain on the ewe with their increased demand for milk, adequate grain is critical. Other feed—such as carrots, beets, cabbage, or cull potatoes fed chopped, and not exceeding 3 pounds a day—are relished and are an excellent aid in the production of milk. Soon before or just after lambing occurs, we put the sheep out to good pasture. We give up to 3 pounds a day of mixed grain in addition to the pasture and taper it down to nothing in six weeks, or upon weaning, depending upon the condition of the lamb(s) and ewe.

Creep feeding is essential to ensure that lambs become established on a diet of high-quality hay and grain as quickly as possible in order to bring them to maximum market weight more rapidly. As mentioned, early in their lives, lambs are not true ruminants; when provided early access to roughage and grain, they are afforded the opportunity to develop this characteristic as rapidly as Mother Nature will permit. Baby lambs should not have to compete with their mothers and other ewes for roughage, as they are sure to come up short at the very time they need the best possible feed provided in an unrestricted manner.

Breeding Rams

Generally, a ram will get by quite well on 3 to 4 pounds of good hay per day. If he seems to be somewhat unthrifty, provide him with a little grain, in the meantime making sure that there are no underlying health problems. For 2 weeks before and during the breeding season, I will give a ram ¼ to ½ pound of grain per day to provide the additional nutrients he needs during this period of increased activity.

Replacement Ewes and First-Year Lambers

I feed our replacement ewes a pound of mixed grain per day all summer long, into the breeding season, to give them the extra feed needed for faster growth and the increased needs associated with pregnancy. Follow any of the programs outlined in Table 3.3 for them, but supply 1 extra pound of mixed grain (or the equivalent), for growth. Your first-year lambers are growing themselves, as well as nourishing a fetus. Monitor them to ensure that they do not become too fat or too run down, and adjust their feed accordingly.

The problem of feeding different rations to your sheep in the winter (e.g., young ewes needing more grain than older ones; ewes needing grain but rams not needing it) is solved by the larger operators by putting sheep into different winter quarters according to what they eat. The small sheep owner wouldn't want the expense of different pens, nor should he have to go to the bother with a small flock. Neither would he want to feed a ram 2 pounds of grain a day when he doesn't need it, or a ewe more than she needs, just because the younger ewes kept with them need that much. It is both expensive and possibly injurious to your ewes to overfeed them. We have resolved this problem by herding all our sheep into the barn together, and closing the gate. We then herd out those who get one ration, keeping inside those that get another. This is difficult at first, as all the sheep want to go out. But do this for one week, and your "dumb" sheep will soon learn. Herd them in, then open the gate; the ones that usually go out will do so, while those that are fed inside will stay in.

Supplenting Commercial Feed

While grass lambs can be raised to butchering weight on grass alone, breeding ewes will generally need other food besides grass or good-quality hay. The numerous feeding programs outlined earlier should afford stock owners a greater opportunity to feed their flock on what they can grow themselves, or what they can purchase cheaply in their local area.

As mentioned earlier, cull potatoes, cabbage, turnips, and carrots are a good supplement to regular feed, and are especially good for milking ewes. In feeding these items, do not exceed 3 pounds per animal. Do not make a change in the amount of grain or other concentrate you feed, but allow your savings to come in decreased hay consumption.

Leftover bakery products from your supermarket can provide you with a very low-cost protein supplement, but careful dietary management is critical to maintaining a properly functioning rumen. Any ruminant is basically a roughage eater, and what we add to their diet is done purely to speed up the growth of the animal to a desirable market weight, increase the production of milk, or improve reproductive efficiency. Without proper and adequate roughage, a diet slanted too much toward stale bakery products can lead to serious digestive disturbances.

MANAGEMENT

Routines

The routines you establish with your sheep will, of course, depend upon your demands and the size of your flock. If you are simply raising grass lambs, you will butcher them all in the fall and be done with sheep until the following spring, when you buy a few more woolly little lambs.

We have a ram and a number of ewes. After lambing in the spring, we keep a few meat lambs and any replacement ewes we might need and sell the rest. This income more than pays for the feed for the ewes over the past winter. Any surplus goes to lower the overall price of our meat and wool.

Auctions are a quick way to sell surplus sheep

or lambs, but generally a higher price can be fetched through private sales. You can usually expect to get 30 to 50 percent more for ewe lambs because of their breeding potential, but check average prices in your area in a market bulletin before selling. Any runts you may have on hand around Christmas and Easter can often be sold at a good profit (quite often higher than larger full-grown sheep) because of the demands placed then by the ethnic markets.

Many sheep owners today gear their lambing time to the Easter lamb market. Develop your fall breeding program to have the majority of the lamb crop ready for that time of year. You need to program the arrival of your lambs so that they will reach market weight at Easter. If you are located within reasonable driving distance of a major metropolitan area, it is entirely feasible to develop an excellent market for your crop. Easter lamb traditionally commands the highest price of the year. Once you have established a clientele, it is possible that you will have a difficult time keeping up with the demand.

A day taken to visit small meat markets in either predominately Italian or Greek neighborhoods may put you in touch with an excellent potential outlet for your lambs.

Another bonus in raising sheep is in their wool and hides. Shearing is carried out in late spring or early summer each year. Depending upon the breed, you may get 18 pounds of wool per sheep, but for the medium-wool breeds the average is 5 to 12. Grass lambs can be sheared before butchering in the fall, and the hides tanned (see Appendix D); or, if you pre-fer, you can tan the hides with the wool intact, for use in making slippers, gloves, rugs, and coats. Unless you plan to be snowed-in all winter, or you have a large family and can delegate work to your children, you probably won't be able to spin all your wool. There are establishments that do custom spinning and dyeing of wool. You can send your wool out for processing and you will get the same amount back, dyed to your specifications—minus a certain percentage, depending on how dirty or full of burrs the wool is. The wool from one lamb can keep you in socks for years.

We generally send all our wool to be spun, and sell it by the skein. We charge far less than the store price for 100 percent wool but still a good deal higher than our cost. This helps to pay off the cost of keeping the flock and possibly generate a little profit.

Some handspinners in your area might well be interested in buying whole fleeces. Also, look into wool pools in your region. These are central points where you can take your wool, tagged and bagged. When the wool is sold, you receive payments minus the cost of running the pool.

Keep signed receipts from the purchasers of the wool and at the end of the year file these at your local Agricultural Stabilization and Conservation Service (ASCS) office so you can participate in the wool incentive payment program. This program is financed out of tariffs on imported wool (not our tax dollars) and pays the seller a premium depending on the wool he has sold and the national average price for wool in this program. The ASCS office near you can provide full details.

Sheep can be beneficial for pasture improvement. As long as you don't allow them to overgraze, they will improve your land. Besides producing rich fertilizer, they also eat about 90 percent of all weeds, so after a few years you should have a more lush, weed-free pasture. Liming and topdressing with poultry or other available or cheap manure should also be done periodically.

Handling

Problems in handling sheep, or retrieving loose animals, will be minimized if you tame them early in life or soon after you buy them. One of their greatest defenses is their timidity, and they will shy away from almost anything, unless given reason not to. If you feed them with grain by hand, petting and talking to them often as you do so, they will begin to lose their natural timidity. While this can be helpful if any get away or during shearing, it can also be a great help during lambing if your ewes feel secure and are unafraid around you.

If your lambs have escaped their pens, a little grain in a dish will usually lure them back. Do not try to reason with them—they will not listen. We have found a snag made out of baling twine to be very effective in catching especially shy sheep. Make a slip knot, attach 10 or 20 feet of twine to it and sprinkle some grain around the loop. Be patient. When the sheep steps into the loop, give a yank and you'll have your stray.

In guiding stubborn sheep to a desired point, gently grab with one hand under the jaw and push with the other from behind. Steer with the head. If you really want to get fancy, you can buy a shepherd's staff from one of the farm supply companies.

Unless you have only a few sheep and you can tell them apart, you will need to mark your flock for identification. This helps immeasurably around lambing time and in your record keeping. A commercial ear tagger can be purchased for easy marking, but with a relatively small flock you can do just as well with a concocted system such as a bell on one, a collar without a bell on another, and no collar on another.

Elsewhere in this book you will find an update on the proposed national animal identification system. In the not-too-distant future, all farm animals will bear some sort of permanent ID. The details are still being worked out at the time of this writing.

Predators

Dogs are by far the most common predators of sheep. Depending upon your part of the country, coyotes may also pose a threat. While your sheep are at the mercy of a predator when tethered, a good fence should prevent animals from molesting your flock. Keep an eye out for any strange dogs in your neighborhood, since they are candidates for sheep killing. Be especially watchful when friends come to visit with their dog. Dogs that have never seen sheep before are naturally quite curious. It usually begins quite innocently ("They're just playing" are famous last words) with the dog playfully running after the sheep. Then he gets all worked up, nips, and nips again, and

draws blood . . . and you know the rest. Prevention is usually your only—and best—recourse. It has been my observation that a single dog is far less liable to molest sheep than two or more working together, and packs pose a real problem. As a livestock owner, in most states you have the right to kill any animal that is wreaking havoc in your herd or flock.

If you lose any sheep to dogs, local dog license funds usually provide compensation for the owner.

On rare occasions, sheep may be lost to bears or other wild animals. Many states provide reimbursement from fish and game funds, providing you can prove loss by producing the carcass—which is often quite a trick because that is presumably what the bear came for in the first place.

A variety of animals have been promoted as guardians for the flock: Donkeys, llamas, and specially bred dogs all have advocates suggesting they do a good job deterring predators bent on thinning your flock without invitation. Investigate some of these possibilities. With flocks of any size, a good herding dog can do much to make herding a far easier job. The Border Collie is the standard for all herding dogs; they love to work, and nothing is more forlorn than a Border Collie with nothing to do. Go to some sheepdog trials and be amazed at what a well trained sheepdog can do. If you are attempting to instill intensity in your own offspring, have them observe a sheep dog at work.

Shearing

Plan to shear in the spring after the weather has settled, when there is no danger of severe weather that would chill the sheep. Wait until there have been enough warm days to bring out the oil in the fleeces; this oil strengthens the wool and also makes it easier to clip. Try to shear before setting your sheep to pasture, as new pasture tends to loosen the bowels, which results in stained fleeces. Proper tagging will, however, cut down on staining if you must put them out.

You should remove all stained pieces and burrs (if possible) before shearing. You'll see that it's wise to dig up any burdocks in your pasture, because a sheep covered with burrs will have a worthless fleece.

Watch someone shear and then try it yourself. You can use either electric or hand clippers. It is important to remove the fleece in one piece, shearing close to the body but avoiding nicking, and avoiding any second cuts. Do not go over the sheep again, as short fibers will result, and the value of the fleece is lower.

If you do not wish to shear your sheep yourself, there are traveling shearers who will do the job. Often they will not come by for just one or two sheep. If your flock is small, arrange for neighbors to gather their sheep with yours in one place to make the shearer's job easier.

Meetings

In most states and areas where sheep are raised, associations of breeders hold periodic informal meetings. Beginners will especially find these useful—and fun. You can usually watch demonstrations of shearing, docking, or castrating, and the almost incredible herding performance of trained sheep dogs, again including the Border Collie. People often take two or three sheep to be sheared on the spot for a nominal fee. Different breeds may be on exhibit; extension service representatives or veterinarians give lectures and hand out helpful informational circulars or bulletins. Frequently, spinners and weavers will also be at work, and people from your locale or region who will buy or trade your fleeces for skeins will be there. These field days are enjoyable opportunities to compare notes and problems with large- and small-scale sheep raisers, or with other backyarders.

BREEDING

If you go into breeding, start slowly. It's a good idea to buy a bred ewe or two the first fall or winter. This frees you from having to purchase a ram and concern as to whether conception has taken place. You might be able to get a good ewe, somewhat advanced in age and not a good competitor in a large flock, but which will do well in your flock. We bought our first ewe in January for $30. She was perhaps 6 years old and doing poorly in the large flock, although she was by no means sickly. No sooner was she in ours than she quickly established herself as the matron of the flock. To this day, she dominates all the other sheep. She has a few more years of lambing left in her, and she has already supplied us with many lambs, and more than paid for herself.

Most of the medium-wool breeds come into heat fairly regularly, beginning in the late summer or early fall and continuing until January, if not bred. The onset of estrus is determined by the photoperiod or the number of hours of daylight, which begin to diminish in late August or early September, at least in the northeast, and this favors the onset of estrus in most breeds of sheep. The trigger for this phenomenon is the effect of light on the pineal gland, which in turn causes changes in other glands that interact, ultimately acting on the ovaries themselves to cause the signs of heat. The heat period (estrus) is when the ewes will permit the rams to mate with them, and its duration is approximately 30 hours. This period will occur every 13 to 19 days, with the average being every 16 days, until the ewe is bred or until the breeding season is over. The Dorset and Tunis and their grades or crosses can breed any time of the year, and caution should be exercised so that they are not bred at inconvenient times. (Inconvenient, that is, for you!) One aid in assuring that the ewe flock will come into heat at about the same time is to take advantage of the so-called ram

effect. Ewes that have been separated from rams will often begin to show signs of estrus if one or more vasectomized rams are introduced into the flock a few days before the projected date for the actual start of the breeding season. This interaction with a male triggers the onset of the reproductive cycle.

In some management schemes, the ram is allowed to run with the rest of the flock the entire year, except before the ewes begin to come into heat in the late summer or early fall. Be extra careful and not risk early pregnancies by removing the ram in plenty of time. In Vermont, the Stevenses pulled their ram in late August and have never had any "mistakes." The other, perhaps less risky, approach is to keep the rams completely separated from the ewe flock in order to completely avoid the possibility of any undesired off-season pregnancies. One ram from yearling age up until 5 years of age can handle 25 to 30 ewes. During breeding, feed the ram one pound of grain per day. After 6 years of age, rams begin to lose libido. If you find yourself with a large number of open ewes, it may be time for a change.

The gestation period for sheep is 144 to 152 days, with an average of 147. To monitor the effectiveness of the breeding program, one method is to have the ram wear a harness that has a crayon marker underneath, so that when the ram mounts the ewe, she is marked. Breeding does not guarantee pregnancy, but it gives the owner a good indicator of the amount of activity going on in the flock. When an estrus cycle has been completed, the ram is fitted with a crayon of a different color and the process repeated. Carefully record the numbers of those ewes covered during each heat cycle.

About 5 months before you want your ewes to lamb, you should move your ram in with the flock. Before doing this, carry out the following procedures: eying (ewes and ram), tagging (ewes), and ringing (ram). Eying involves clipping the wool from around the face and eyes in closed-faced breeds. This improves the eyesight and prevents eye irritation from the wool and any seeds and dirt carried in it. Tagging is clipping the wool around the dock of the ewes. This promotes cleanliness and permits the ram to couple with the ewes more easily. Some people shear a ram prior to the breeding season, but in a cold climate such as Vermont we do not do this because the ram will not be able to grow a full fleece back before the onset of winter.

Ringing is an acceptable substitute, and consists of clipping wool from the neck and from the belly around the penis. This makes it easier for the ram to make proper contact with the ewe during mating. Excessive heat may affect the potency of your ram adversely, and if breeding is to be carried out in very hot months, complete shearing may be necessary to ensure full potency. Your sheep should not be wormed during the month before they are bred.

Two weeks before the start of the breeding season, you should flush the ewe flock. Flushing is the shepherd's term for increasing the level of nutrition in the ewe flock in an attempt to raise the level of conception. Research has shown that this increase in feed will promote more eggs to be shed per ovulation, thus increasing the likelihood of a successful breeding.

It is very difficult to tell when ewes are in heat. Sometimes they will begin mounting each other and their vulvas may be red and swollen, but this swelling is not nearly as pronounced as in pigs. You can usually tell if they allow the ram to mount them. In most cases, a ewe will not be receptive to the ram unless she is in heat. If you can see your ram breed with your ewes, you calculate approximate lambing time from that date—assuming it was a successful insemination. Or you can just leave the ram in and assume all your ewes are bred, and have the lambs come as a surprise the following spring.

A better system in which you can be reasonably sure of the date of lambing is to equip the ram with a marker on his chest so he marks the back of the ewe when he mounts her. You can buy a harness with a crayon attached from farm stores, or you can mix a marking paste to apply to the ram's chest once a week or so. A simple paste can be made by mixing mineral or linseed oil (do not use motor oil) and any available nontoxic dye such as Venetian red or lampblack. Change the color of the crayon or paste every 16 days, using in progression yellow, red, and then blue or black. (They are used in this order because they cover the previous color.)

The procedure is as follows: Equip your ram with the first color. Within 16 days (the average heat cycle), all the ewes will come into heat and should be mounted by the ram. They will then be marked by the first color. The day you notice the mark on the ewe is approximately 21 weeks from their lambing time—tentatively. After 16 days, change to the next color. If the ewes are not marked with this color, it means they are already bred. Those ewes that are bred will not come back into heat, and they will not accept the ram. These ewes should lamb 21 weeks from the day of the first marking. Any ewes marked with the second color should lamb 21 weeks from that day. The color will be changed again after that 16-day period, and so on. In small flocks, all ewes should be bred in their first or second heat after being with the ram. If none of the ewes is bred, your problem is most assuredly the ram. If a few ewes are not bred, they may be too old, in poor health, or sterile. They should be considered for mutton, or for sale.

While sheep in commercial flocks are bred for only 5 years, those in small flocks generally receive such good care that they can lamb successfully for as long as 10 years. As a sheep gets older, its teeth deteriorate, hampering food intake, and in such a case it may not breed successfully, or may not be able to nurse the lamb, or it may be so physically taxed that it will die during pregnancy. Keep an eye out for a ewe having trouble. Sell, butcher, or keep her as a pet, living out her twilight years without the drain of children (but still supplying wool, of course).

There is some disagreement as to how young you can begin breeding a ewe. Some say they shouldn't be bred until their second year; others recommend breeding in their first year, to lamb at the age of 14 months or so. Recent studies have shown that ewes bred as lambs will produce a greater lamb crop during their breeding life than those bred as

> ### LAMBING:
> ### WHAT TO HAVE HANDY
>
> A clean pail
> Heat lamp
> Petroleum jelly, mineral oil
> Soap
> Soft, heavy string
> A lambing loop (nylon twine)
> Uterine capsules
> Dry rags, towels
> Ammonia or smelling salts
> A helper

yearlings. Ewe lambs should be at least 9 months of age and well-grown if they are to be bred.

Ewes that are bred at an early age must receive a diet adequate to meet the needs of the developing lamb as well as her own needs for growth and maintenance. These ewes must have particular attention at lambing as they are more inclined to have difficulties than older sheep.

These young ewes also are more inclined to refuse to accept their lambs, thus presenting the owner with the additional problem of possibly bottle-feeding the young.

A week or two before you believe lambs are due, check to ensure that the ewe's udder and teats are free from wool. If not, clip that area and around the eyes (tagging again) if necessary.

On the day of lambing, a ewe may exhibit one or more of the following signs: She will be off her feed, restless, isolated from the rest of the flock, or dig or paw at the ground or bedding while circling a particular spot. She will appear sunken in front of the hips; the vulva may be red and enlarged; and the udder will be substantially larger and the teats coated with a waxlike film.

When you notice any sign of labor, watch but leave her alone and don't upset her by constantly peeking and feeling her stomach. Sheep may need help at times, but they are not nearly as prone to problems as they are reputed to be. First-lambers tend to have more problems, but if your flock has been well fed and exercised, you shouldn't have any major problems. A friend of ours who has the largest flock in Vermont had 300 lambings one spring, and had to assist only in one.

Do not leave pails of water where lambing is to take place, because it is possible that lambs will be dropped into a pail and drown. Offer water to the ewes periodically, and then remove the buckets.

The normal position at birth is the backbone of the lamb toward the back of the ewe, nose presented first with front feet presented alongside it. If afterbirth is passed soon after the birth, generally there will be no more lambs. If afterbirth doesn't come and the ewe continues to show signs of labor, paying little attention to the lamb, another will probably follow. In some instances, three lambs (or even more) may be born, but you needn't bother yourself about such occurrences.

If the lambs are born out on pasture, you needn't worry, as long as they seem healthy and the mothers clean them off and allow them to nurse. If born inside, cut the umbilical cord off at 4 inches and apply 7 percent tincture of iodine to prevent infection.

If conditions are crowded, place the ewe and her lamb in a lambing pen for a couple of days. You can put the ewe there before lambing if she is obviously ready to drop the lamb. Make sure she can see the rest of the flock and doesn't get upset. This allows the ewe and her lamb to get acquainted without danger of losing each other—most important for yearling and ewe lambs.

In moving the new lamb to the pen, grab it gently under the chest and back it into the pen, keeping it under the ewe's nose. For at least the first few days, the ewe knows her lamb by smell only, and if she or it wanders away and it mixes with other sheep, she may disown it.

In chilly weather some burlap or blankets hung around the outside of the pen will help keep out drafts. A heat lamp should be hung in the corner of the pen so that the lamb can get close to it by choice. To prevent ewe or lamb from getting burned, hang it 3 feet above the floor. Do not use it for more than 3 days, because the lamb might get chilled when removed from the pen. If all goes well, the lamb and ewe can be let out of the pen in 2 days and allowed to run with the flock. Follow the feeding instructions listed earlier and you will minimize your problems. You can wean the lamb as early as 4 weeks or let it wean itself naturally. We generally keep all our sheep together, and in time the ewe will dry up and not allow the persistent lamb to nurse. As the ewe dries up, or as an aid to help her, slowly taper off her grain supplement until her feed routine is back to normal.

Problems During Lambing

Sometimes 3 to 4 hours are required during lambing. Examine the ewe a half hour after the first water bag is passed and note her progress. Have all those helping wash their hands thoroughly. If you enter the ewe with your hand, coat it first with a lubricant such as petroleum jelly. Make sure your fingernails are short and smooth, and if possible bend your fingers at the knuckle before inserting them to avoid scratching. Lay the ewe on her right side and probe with your hand, trying to determine if the lamb is in the proper position. Ideally, its nose should be between its feet. (Now you can see the benefit of having your ewes tame and used to your presence.)

A few hints: To distinguish between the front and hind leg, feel above the knee. The hind leg will have a prominent tendon; the front leg, muscle. If the hooves of the lamb are pointed down (toward the backbone of the ewe), they are probably front legs coming in the normal position. If they are pointed up, they are probably the hind legs and the lamb is being presented backwards.

The most common difficulty is when *one or both* front feet are folded underneath the lamb. In the case of one foot being folded back, tie a soft string around the presented leg (so it will not be "lost" in this procedure) with plenty of string protruding. Then push the lamb back slowly and gently. By sliding your hand down the neck, you can locate the leg that is turned back and straighten it out. The lamb is then in the normal delivery position and should be delivered without further problems. If the ewe is too fatigued and cannot pass the lamb unaided, use nylon twine to make a noose that is placed behind the ears and then into the mouth. String tied around the feet can be pulled along with the loop and the lamb should come. You can pull quite hard, surprisingly, without hurting the lamb. However, always pull *with* the ewe's contractions.

In the case of *both front legs being doubled back,* try the same procedure as outlined above. If that does not work, place a loop of string around the lamb's head, with the knot in its mouth. Elevate the ewe's hindquarters and gently force the lamb back until you can run your hand down the neck and flip the legs through the pelvic arch. If the head has not slipped back through the arch, start it through before you start the legs through. In all of the above operations, lubricant applied to the vaginal wall will make it easier to maneuver the lamb into position.

Often, in the case of hard labor, the head of the lamb swells and makes lambing even more difficult. If the head is turned to either the right or the left, and you find it impossible to keep it in position to work it into the birth canal, fashion a loop of twine or nylon string as suggested above. This can then be used to move the lamb along.

In the breech position, the lamb is backward, and the rear legs are under the belly of the lamb and pointing toward the front legs. The tail is often the only thing that can be identified. On examining the lamb, the rear legs sometimes are located only with difficulty, so slow, careful, systematic exploration is critical. Once the hocks are identified, they are grasped one at a time, with the thumb and forefinger.

Slow, gentle traction is applied as the leg is brought backward. If you are successful in accomplishing this maneuver, it is usually possible to slide your hand downward and forward until you can grasp the hoof. With the hoof in your hand, you can turn it inward and, at the same time pull it backward. The same procedure is repeated on the opposite leg and the lamb is then delivered by applying firm traction.

Sometimes you will be presented with an abnormally large lamb. If it is in the normal position, work the legs forward a bit and pull them gently while using the other hand to pull the head from side to side with the fingers positioned behind the ears. A helper can stretch the top of the vulva to ease the operation.

When you are faced with an obstetrical procedure, do not hesitate to call for professional help. if you feel you are in over your head. Don't belabor a difficult delivery; instead, recognize your own limitations and call for assistance, before the situation gets out of control. A dead lamb, or ewe, is a poor reward for failing to recognize a situation that is beyond your abilities.

Problems After Lambing

The lamb, once delivered, often needs assistance in starting to breathe. Time is of the essence, and the necessary supplies should be nearby in the lambing pen. An old, clean facecloth or piece of towel works well to remove the excess mucus from the nostrils and the rear of the mouth.

The next step should be to give the youngster a brisk rubdown with a towel, especially over the chest, drying as well as attempting to stimulate contractions of the ribs. Don't be too rough—or too tender.

Next, grasp the newborn by the hind legs, elevate it so the front hooves are just off the ground, and swing the lamb back and forth in a gentle arc. It is amazing how vigorously one can swing the lamb without doing any apparent harm. This will help to ensure that mucus and other fluids in the back of the mouth and in the nostrils are expelled. Oftentimes this is enough to stimulate the newborn's breathing.

If breathing has not started, lay the lamb on the ground and pour a small amount of cold water into one of its ears. This simple procedure will often stimulate the breathing reflex when other attempts have failed.

Another source of stimulation is to have either household ammonia or smelling salts on hand. Pour a small amount of either solution on a piece of cotton and hold it close to the lamb's nostrils. This can have a powerful effect on stimulating breathing.

If a lamb is chilled, it should be immersed in a pail of water heated to 90 to 100 degrees Fahrenheit. Remember that it has been in its own incubator with an ambient temperature of 102 to 103 degrees. Leave all but its head in the water for 10 to 15 minutes, then towel it off quickly and place it under a heat lamp. The downside of this method is that much of the lamb's odor will be washed off and the ewe may disown it.

It is essential that the lamb get an adequate amount of colostrum within a short time after birth.

Colostrum supplies the critical antibodies that ensure that the lamb will be protected against the many diseases to which it is susceptible in the first days of its life. If these antibodies are not supplied in the first 12 to 18 hours of life, additional colostrum later will be of little or no value for immunizing. The reason for this is a process called closure, whereby the intestine can absorb the colostral antibodies for only that short period of time after birth.

Nursing

If the lamb does not nurse within an hour, be patient. Allow the lamb to gain its strength slowly. You may have to aid it. Put some milk on the end of the teat and push the lamb's mouth to it, or strip some milk into its mouth. Tickling the underside of the lamb's tail while it has the teat in its mouth may help.

If a mother doesn't clean her lamb off, place it by her head to encourage her. If she still doesn't do so, dry it yourself with a clean rag or towel. If she won't accept it or refuses to let it nurse, you will have to assist the lamb. Often the reason a ewe disowns a lamb is that she is first-lamber bewildered by the whole lambing procedure, is in a rundown condition, or has udder problems. In the latter case, pressure and discomfort in her udder may make it uncomfortable for her to have her lamb nurse. Reducing the feed intake the day after lambing will go a long way toward lessening such problems. Try to strip some milk out of the ewe to help lessen the pressure and to allow you to evaluate the color, consistency, and odor. Apply hot packs and keep the lamb nursing if she will allow it.

If she still refuses the lamb, then you can only counter with perseverance and patience. Don't give up. On one occasion, we worked for 5 days with a ewe before she finally allowed the lamb to nurse. You must try to get some colostrum into the lamb, either by holding the ewe and letting him nurse or by stripping some from her teat and giving it to him with a dropper. Rubbing some afterbirth or some of the ewe's milk on him will often encourage her to accept him.

A little grain sprinkled on the lamb may make her more receptive. Something to distract and worry her, such as a dog tied within her view, will get her mind off her resistance to nursing. Hold her on her back and let the lamb milk, or face her into a corner so she can't move and hold her back legs so she can't kick him away.

Keep at it. Remember, each rebuff discourages the lamb even more; however, each successful nursing heightens the lamb's confidence and begins to make the ewe more accepting. The thrill of seeing a ewe accept a lamb after 5 days is well worth the effort. If these measures fail, give her another chance the next year. Should she repeat the performance, cull her.

Feeding the Young

Sometimes the lamb refuses the mother. The author of this book happened to be away when one of his older ewes lambed. Since one of our own pregnant ewes was being boarded on the author's farm and was due any day, we had stopped by to check and stayed to give the stock and house sitter a hand.

The ewe lamb looked fine, but she wouldn't nurse,

even though the mother seemed calm and cooperative. We were also puzzled by no signs of the afterbirth, and wondered if a twin was on the way, but after 4 hours we gave up hope of that.

Aware of the infant's immediate need for colostrum, we tried every device we knew and had used before. The infant would suck on a finger but not a teat. The editor managed (for the first time) to strip the ewe while one of us held the lamb's mouth open, but that wasn't really very effective. We called the vet to make sure we were proceeding in the right direction. He told us the ewe could have eaten the afterbirth—though we could reach in and find out if we wanted to be sure (we didn't)—and that we should give the lamb a couple of ounces of the ewe's milk every hour or so in a baby's bottle with a nipple.

We did so, after borrowing a bottle from a neighbor. *Keep one on hand for just such emergencies.* The nipple hole had to be considerably enlarged, and although the lamb did not swallow much, it was evidently enough to prompt her to start nursing normally as soon as we put her back in the pen with her mother. Lesson: even rank amateurs can rise to the challenge of lambing!

If a ewe will not nurse her lamb, or in the case of an orphaned lamb, you have two options: raise the lamb by yourself or try to "graft" it onto another ewe that has lost her lamb. Graft only in the case of a mother whose lamb has died. Do not attempt it on a ewe who has lost one of a set of twins. Dip the lamb in warm water to wash away its odor, dry it, and rub it with the afterbirth of the dead lamb. Let the ewe lick it and clean it off.

Skinning the dead lamb and tying the hide on the lamb may also help. Restraining the mother and applying the same methods as with a real mother who rejects her own should produce results. Keep them confined in a lambing pen for a number of days and make sure she lets it nurse and doesn't injure it.

Orphans

If a lamb must be raised without a mother, foster or otherwise, you can feed it by bottle or use a pail with a nipple on it, as you would with calves (see Table 3.5). I recommend using the pail for a number of reasons: First, the lambs can feed themselves and you will not spend nearly as much time as with bottle-feeding; second, a bottle lamb becomes a cosset, which is the type of lamb that followed Mary to school one day. A cosset becomes as dependent on the milker (you!) as it would its mother, and goes bananas when you leave it. The bleating will drive you crazy and when it comes time to butcher—well, could you really do it?

It is very important to get some colostrum into the lamb as soon as possible after birth. Cow's milk and conventional milk replacers are not good for raising lambs because they are too low in fat. A minimum of 30 percent fat and 25 percent milk protein is necessary. Manna Pro Lama, a Calf-Manna milk replacer, is specially formulated for lambs. You may have to use a dropper to feed for the first few days, but get it on the pail as soon as possible.

Lambs do better on cold milk than warm, because it will not sour so quickly and they will drink it for more frequent and smaller feedings.

Table 3.5: Feeding Schedule for Orphaned Lambs		
AGE	**AMOUNT**	**FREQUENCY**
1–3 days	2–3 tbsp.	every 2 hours
4–5 days	3–4 tbsp.	5 times a day
6–7 days	1/2 cup	4 times a day
2 weeks, inclusive	3/4 cup	4 times a day
3 weeks to weaning (3 months)	until full, 3 times a day	

This is also easier for you. If you are lucky enough to feed it by the pail (and freezing is not a problem), let the lamb feed free choice from the pail. Be sure to change the milk daily, and wash and disinfect the bucket and nipple. If the lamb won't accept cold milk, warm it at first and gradually decrease the temperature. If the lamb can be put out to pasture soon, it will help, as will creep-feeding.

Pinning, or the sticking of the young lamb's tail to its anus, can be a problem for the first few days and, in extreme cases, may even cause death. Check your young lambs by picking up their tails and loosening them, cleaning off the excrement if necessary. When they are a few days old, the bowel movement becomes firmer and pinning is no longer a problem.

Castrating and Docking

You should castrate some rams, but not those that you will market for meat when they're 6 months old or younger or those you may keep or sell for breeding. Rams grow faster than castrated males, while the latter grow faster than the ewes.

Castrating should be done as soon as the testicles descend into the scrotum. An emasculatome may be used, especially during fly season as there is no open incision left using this method. The emasculatome, a pincer instrument, cuts off the blood supply to the testicles and they become nonfunctional and soon atrophy.

An Elastrator works similarly, stretching a rubber band around the neck of the scrotum. When the Elastrator is removed, the rubber band tightens and cuts off the blood supply and the testicles atrophy.

Finally, a sharp knife can be used. Because this creates an open wound, you should work in a clean place to avoid infection. One person should seat the little ram on its rump, then hold a front and rear leg in each hand. The second person cuts off the lower third of the scrotum, squeezes the testicles from the scrotum, then snips them off. The wound should be sprayed with an antiseptic.

Docking is recommended. Lambs, both male and female, should have their tails cut off, or docked, in the first week of their lives. It makes mating easier for ewes, results in cleaner sheep, lessens the chance of maggot infestations, and does not get in the way at lambing.

Any of these three tools can also be used for docking. Whichever method is used, the cut should be at the third joint, about 1½ or 2 inches from the body, the tip of the tail should be just above the top of the anus when properly done.

If using a knife, locate the place to cut, then push the skin toward the body so there will be extra skin to cover the stub. Twist the tail a quarter turn so the blood vessels will be cut diagonally and lessen the chance of severe bleeding. After cutting, pinch and hold the end to stop bleeding.

The Elastrator may be used. Using this system results in the tip of the tail dying off slowly, immediately raising the likelihood of infection.

The emasculator will sever the tail of the youngest lambs, but with those that are a few days older, a combination of that tool and a knife may have to be used, cutting the tail outside the jaws of the tool. There is very little bleeding using this method, since the blood vessels are crushed.

If you have any reservations about trying any of these operations, watch them done by an experienced person before you attempt them.

HEALTH

Worms are the most common affliction of sheep, and they're best controlled by pasture rotation, together with a good worming program. If you change the sheep's pasture every 6 weeks, the worms' life cycle will be broken. We generally worm our sheep every 6 weeks, or every time we rotate pastures, whichever comes sooner, and every 6 to 8 weeks in winter.

The deworming preparations available today afford a wide range of choices, most of them of excellent quality. These medications are marketed under a variety of trade names, but there are a few major families of drugs that should be remembered at the time of purchase. These are the benzinidoles, the pro-benzimedazoles, the pipantel group, and Ivermectin. An ideal worm-control program incorporates a number of each of the major groups into the treatment and prevention program. One common routine is to use a member of one family of dewormers perhaps three times in a row, then switch to another family for the next three times to avoid the parasites building up resistance to one particular type of anthelmintic given too frequently. The interval between treatments is dependent on the situation on the farm, but an average interval would be 6 to 8 weeks.

One way to keep track of the worm load is to have a fecal exam done on a regular basis. Pick up samples of fresh pellets from around the barnyard and take them to your veterinarian for microscopic examination. This will give you a good idea of the effectiveness of a particular wormer and whether to continue using it in your operation.

In young lambs exhibiting bloody diarrhea, coccidiosis should be considered as a possible cause. When dealing with illness, it is always possible the animal has more than one problem. Finding one possible cause of a problem doesn't mean there aren't others. So it is with coccidiosis. Often one or more additional intestinal parasites can be aggravating the illness, so an already acute condition is made more serious by their presence.

Many good medications are available for the

prevention and treatment of coccidiosis. In attempting to diagnose any intestinal problem where diarrhea is one of the symptoms, you should take a sample of the manure to your veterinarian for an accurate diagnosis. Livestock owners waste money by attempting to "shotgun" treat a problem. All the *Kaopectate* in the world won't cure a case of diarrhea caused by coccidia, even though it may give temporary relief. Do not make the mistake of waiting until your patient is too far gone before getting a little expert advice.

Sheep are infested with a wide range of intestinal worms with formidable names and intricate life cycles. For our purposes, we can say that sheep are susceptible to roundworms, tapeworms, whipworms, Strongyloides, and a host of others. These worms tend to set up housekeeping at rather specific points along the intestinal tract from the stomach to the small intestine, and all the way back to the large bowel.

Limited acreage available for pasture makes control measures somewhat difficult. Sheep are close grazers, and, as such, are particularly prone to picking up new infestations of parasites. Many of these worms spend part of their life cycle in or on the soil and vegetation, and sheep readily pick up new parasite loads because of their grazing habits. Most of us are locked into limited acreage and don't have the luxury of vast expanses of pasture so that we can rotate in such a way as to lessen the chance of reinfestation. Since we must survive in the situation in which we find ourselves, we must adapt to the limitations of our own farms.

Ideally, most of us would love to have a clean pasture lot for our recently wormed flock to lessen the possibility of worm reinfestation. We should have a minimum of two lots, so that we can alternate on some sort of a regular basis. Anything beyond this is a bonus.

Several species of intestinal parasites cause a variety of problems in sheep. Of all of the domestic livestock, the sheep is probably the most susceptible to the adverse effects of worms and is in particular need of all possible control programs.

External Parasites

External parasites usually lead to excessive itching, a generally unthrifty appearance, and, with sucking lice, severe anemia if allowed to go unchecked.

Lice: There are two varieties of lice: biting and sucking. Both can be treated by dipping in an approved dip. There are strict federal mandates on the use of various insecticides used on farm animals. If you are unsure about a particular product, check with your veterinarian before using it. A list of currently approved products appears in the Appendix.

Often lice are noticed in the winter when the use of a dip is not possible. In this situation, use one of a variety of dusts that will kill these miserable pests. Animals seem to use dust bags readily if they are made available. These are porous fabric bags that can be filled with an approved dust and hung in an area where the sheep can rub against them, thus getting dust into their fleece.

Ivermectin, which can be given by injection, is particularly effective against sucking lice.

Mange: Mange is caused by a variety of small mites that burrow in the superficial layers of the skin, causing intense irritation. The type of mange most common in this country is chorioptic mange. Most often it is seen on the legs and between the toes of infected animals. Lime sulfur dips are one effective means of controlling this parasite. Toxophene is another preparation that gives good results. Again, I caution you to keep current on changes in regulations governing the use of chemicals on sheep, as well as other animals whose meat or products might find their way into the human food chain.

Maggots: Maggots are one of the most disagreeable parasites to infest animals. They develop from the eggs of a fly that is attracted to moist, foul-smelling areas resulting from wounds or persistent diarrhea, especially areas where there is thick wool. These miserable creatures feed on debris and dead flesh, and can have a profound effect on an animal.

In large numbers, they can cause an animal to go into a shock-like condition. Unless removed, they can threaten the animal's life. Every sheep raiser should be aware of their existence and, during fly season, be especially vigilant in checking animals that have been injured or have had surgery.

Maggots should be removed by hand, a job that no one approaches with enthusiasm, but one that must be done. Hydrogen peroxide applied full strength to the affected areas is useful in getting them up and moving. Once all the offenders have been removed, the area should be treated as an open wound, kept clean, and covered with a repellent to prevent future episodes.

A product that works well to both kill maggots and repel the flies is available from your veterinarian.

Sheep Tick: Sheep keds (sheep tick) cause skin irritation by sucking blood from the skin. When sheep are infested with keds, the wool becomes ragged and discolored. Dusting, dipping, or spraying the sheep will control these insects.

Worms: The natural resistance sheep possess against worms fluctuates somewhat; the ewe seems to have less resistance against worms just before and just after lambing. Worming should be done 2 weeks before and just after lambing.

Flukes: Flukes are a significant sheep parasite in many parts of the United States and throughout the world. These parasites spend part of their life cycle in snails, so one way of controlling flukes is killing the snail host. If this cycle can be broken, a fluke control program can be considered to be successful.

Flukes enter the sheep's liver and, given time, will destroy the function of this most important organ. A number of effective drugs will control flukes. Consult a veterinarian for the specifics of a control program.

Trimming Hooves

Trimming hooves once a year, usually when sheep are sheared, is important in good flock management. Set the sheep on its rump with its head resting on your left thigh. This position will normally quiet the sheep. Do not hold it in this position too long, or breathing may be hampered. Trim the excess hoof off the sides so it is nearly level with the sole. Do not cut too deeply or lameness may result. If the foot is soft, a penknife can be used for trimming; if it is harder, pruning shears should be used.

BUTCHERING

While the details of butchering sheep are beyond the scope of this book, they are one of the easiest "large" animals to butcher. After watching it done once or twice, and with the help of a good guide, most stock owners can easily do it themselves in an hour or less.

A lamb will dress out at about half of its live weight. A grass lamb, with minor variations due to breed, will furnish 30 to 40 pounds of meat when butchered in the fall. A lamb, technically, is a sheep less than a year old; a yearling is aged 1 year to 18 months; and mutton is any sheep over 18 months of age. The older sheep get (or any animal, for that matter), the stronger the meat tastes. However, while mutton may be a bit coarser, the popular belief that it is no good is a total fallacy, as far as I'm concerned. Commercial mutton, maybe. But home-grown mutton or lamb is another slice of meat altogether.

CHAPTER 4

Goats

The dairy goat is the ideal animal for the family that has reached its limit in paying the climbing prices for milk and other dairy products. A good doe will produce 2 to 6 quarts of milk for up to 200 days per year. By having two goats and breeding them at different times, you can have at least one of the goats milking all year and still have some surplus for sale, or to feed to other stock.

Those of you who are not familiar with goats and goat's milk should realize that you are in the minority—more people in the world drink goat's milk than cow's milk. Goat's milk is, in my opinion, indistinguishable from cow's milk (maybe a bit better, to quote some "unbiased" sources) and even more digestible. And don't forget other dairy products, cheese, and butter, as well as leather from the hides, and chevon (goat meat). All this for a fraction of the cost and space requirements of keeping a cow.

Being ruminants, goats are also ideal for our protein conversion purposes. While they have the same ability as sheep to convert low-protein feed into high-protein products, they are perhaps even more effi-cient feeders because they are less picky about what they consume. They are truly a browsing animal—like deer—and while they will consume grain and hay, they also delight in shrubs, weeds, saplings, bark, hardhack, and the like. They would rate number one in this book in feed efficiency, and they also rate number one in personality. If you've ever had a goat, you know what I mean; and if you're getting one, you'll find out.

BREEDS

Five recognized breeds of goats predominate in this country, and you will be considering either a purebred or some cross of one or more of these breeds.

The first three, referred to as the Swiss type

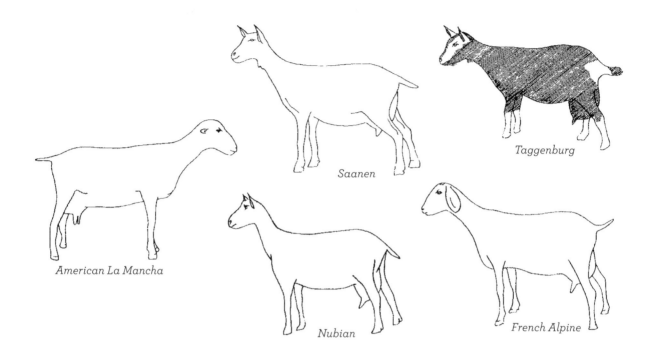

FIG. 4.1: THE FIVE RECOGNIZED BREEDS OF GOATS

because of their similar conformation and origin, are the French Alpine, Saanen, and Toggenburg (Fig. 4.1).

French Alpine: Similar to the other Swiss breeds in conformation, the French Alpine can be almost any color combination. It has prominent eyes, erect ears, a straight (as opposed to dished) face, and a graceful appearance.

Saanen: Similar to the Alpine but with a dished face. Color is pure white to cream. Saanens are the high-

est milk producers in the group, but their milk has a relatively low fat content, a minus for those who wish to make a lot of cheese and butter, but a plus for those with dietary considerations.

Toggenburg: The smallest of the Swiss type, with the same basic conformation as the Saanen. Toggenburgs are brown to light fawn in color, with white markings in the form of stripes on the face, white ears with a dark spot in the middle, and white on the legs, near the rump and around the wattles.

The two other major breeds are American La Mancha and Nubian.

American La Mancha: A relatively new breed that has almost no ear flaps. La Manchas have straight faces and various color combinations. They are quite calm and good milkers.

Nubian: A popular breed with long, droopy ears and a Roman nose. Nubians have a variety of colors and combinations. They are most noted for the high butterfat content of their milk, but the lowest volume of production, on average.

PURCHASE

Within each breed you will see different types of breeding: purebred, American, recorded grade, crossbred, and unrecorded grade. A *purebred* is a goat whose parents are both registered and of the same breed. An *American* is a goat that is a result of grading up, by breeding three successive generations of a goat to purebreds of one breed. The result after three generations of breeding will be seven-eighths of that purebred and be termed an American (e.g., a grade goat bred with a purebred will produce a kid that is half purebred; the next breeding will produce a three-fourths purebred; the final is a seven-eighths purebred, or "American"). The American La Mancha breed came about by breeding a short-eared grade breed with purebred stock. A *recorded grade* is a doe (bucks of this type cannot be recorded) with one parent of unknown or

mixed breed and the other a registered purebred. A *crossbreed* is the result of breeding two purebred parents of different breeds; and an *unrecorded grade* is of unknown or unrecorded parentage.

As far as breeds go, there is not enough difference between one and another to single out any as superior. If you want more byproducts such as cheese or butter, you might try a Nubian; but any breed should fit your needs. By far your chief concern should be: Is the breed of your choice popular in your locale? You may have the handsomest purebred Nubian, but if there are no others around, you'll have a problem with breeding. Check with any goat clubs in your area to find out which are the most common breeds in your area.

What type of animal should you buy: purebred, cross, grade? Should you buy a kid, yearling, or older milking doe? There are no pat answers to these questions; it will depend on you. A purebred doe will not necessarily produce more than an unrecorded grade, and she will cost more. However, with recorded animals you have records of their parentage and perhaps milking records. With any unrecorded animal you are gambling. You might get some hints as to future production by examining conformation, but records are the best source.

Purebred goats will also have more valuable offspring. You'll always have kids to sell, and, if they're purebred, the chances are you'll find a better market, both for does and for breeding bucks, and they will fetch a higher price.

If you buy a 6-week-old kid, you may have to wait up to a year before you get milk. A yearling will cost more but will be ready to give milk. An older doe will cost somewhere between a kid and a yearling, but will have a less productive life ahead, and is more of a gamble for all but the most experienced buyers.

Unless you have records, determining the exact age of a goat is difficult, but you can check the teeth the same way you do with sheep. It's probably best not to get a buck to start with (more about bucks in the section on *Breeding*). When you are looking for your goat, if possible take an expert with you, and do not be in a rush. Shop around; again, going to shows or fairs and noticing the top-rated animals will help you learn good points to look out for.

Body

While we look for blocky, heavy conformation in meat animals, dairy animals should have a more angular conformation: thin thighs, prominent hip bones, and a comparatively "lean", though not sickly, appearance. Any thicknesses of the body, such as a short neck or any fatty areas, are indicators of poor milk producers. A good dairy goat should be rugged-looking, with a broad chest and large girth and with front feet set wide apart. The back should be straight (not arched or swayed) and there should be a gradual slope from the hip bones to the tail. A wide barrel and ribcage indicate good food capacity and a potential for large litters. The ribs should be far enough apart to slip a finger in between. The skin should be smooth, thin, and supple.

Feet and Legs: The goat should stand erect and move about with ease. Check hooves for foot rot and poorly trimmed feet.

Head: The goat should have bright eyes and an alert appearance. Check the ears for any sores or scabs. Avoid a doe with any lumps or malformations of the jaw. Check the teeth to help in calculating age, and reject a doe with missing, excessively worn, or broken teeth.

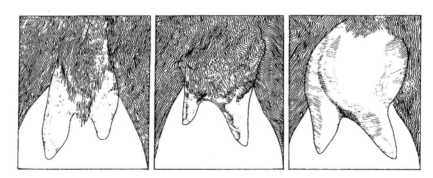

FIG. 4.2: CHECK THE UDDER FOR A HEALTHY FORMATION

Udder: The most important part of any dairy animal. Be certain that the udder is well formed and free from any injuries, scars, or extra teats (Fig. 4.2). It should have a lot of capacity and be held high enough to avoid potential injury from hitting any objects. Look for one that is soft, with no hard spots in the udder proper, or the teats (the udder will be firm if the doe has not yet been milked). While the udder will never be perfectly balanced, avoid one that is terribly unbalanced, and avoid a doe with overly large teats or ones that are pointed sideways. ("If they look you in the eye, don't buy.") Ideal teats are the right size for ease of milking, and will be tilted slightly forward.

If you know how, milk the animal to determine capacity and temperament. See if the seller has any milking records of this doe or her ancestors. Records beyond the second generation, no matter how impressive, are of little use and can be misleading. Look for a long milking cycle and high production. Producing between 3,500 and 4,500 pounds per year is a good average. Detailed records also show that the seller has an interest in his flock and strives to improve it.

HOUSING

Housing for goats is almost identical to that for sheep, with a few exceptions. Goats are not as hardy as sheep and need to be more protected from drafts in winter; they are better at getting through fences, so more care is needed. By reading the section on housing in the chapter on sheep, you'll get a good idea of what you will need, with the following modifications, additions, or deletions.

Pasturing

It has long been recognized that goats are browsers, often preferring to consume plant species that other livestock would consider unappetizing. Not only do they seem to like this sort of marginal vegetation, they actually seem to thrive on it. Of all the plants that one might consider unappetizing, multiflora rose would surely rank near the top. Those of you who have literally wrestled with this noxious species will appreciate my evaluation of this plant. In the 1950s, the extension service advocated its use as natural fencing, its aggressive growth habit and thorns making a natural barrier that most domestic livestock would not attempt to penetrate. The thinking at that time was that the cost of fencing, always expensive, could be virtually eliminated by substituting this most invasive species. It appears that goats are not at all phased by it, and their browsing habits seem well adapted to eating it. Having fought the good fight against this aggressor for many years, if I were challenged again I would definitely enlist the aid of a few goats to assist me in the eradication project.

Tethering: Goats, like sheep, can be tethered but they don't take to it as well, thus extra vigilance is needed. While goats are ruminants like sheep, they browse more like deer. They not only need grasses but a wide variety of feed: plants, small trees, shrubs. Tethering may affect their feed intake, because the limited range of tethering allows access to all the browse they want and need.

Fencing: Goats are much harder to confine than sheep and may be one of the most difficult species to keep where you want them. Electric fence can be utilized, as with sheep, but use three wires, one each at 10, 20, and 40 inches. You'll need a strong charger and may have to train your goats to adjust to the fence. This probably is not the best method.

The USDA recommendation of a 5-foot fence for sheep should also be adequate for goats. They like to climb on the fence and if it is loose they can, in time, pull it down. Use posts at frequent intervals and get the fence as tight as possible. If this still won't contain them, you may have to resort to the absurd-sounding solution of a fence within a fence. Either use a strand of barbed wire or, even better, electric fence set 10 inches high and 10 inches away from the stock wire. A strand of electric wire 10 inches from the stock wire and a foot or two from the top may serve to discourage

Make sure your goats have shade while pasturing.

climbing. You may not need all of this, so don't let it scare you off.

Don't have any young trees within the fenced area or they will soon be devoured. It is advisable that trees be located out of the sight of your goats, lest they become too great a temptation. Goats, like sheep, are sensitive to certain poisonous plants: buttercup, bracken fern, cowslips, false hellebore, Dutchman's breeches, water hemlock, mountain laurel, sheep laurel, sneeze weed, white snakeroot, wilted or dry wild cherry leaves, dry oak leaves, rhododendron, hemlock, azalea, milkweed and locoweed, yew, foxglove, delphinium, lobelia, and lily-of-the-valley. This is not a complete list. Your state extension agent should be able to furnish you with a complete list of poisonous plants found in your region.

Your goats need a shady area. Gates should be tightly secured because goats are quite adept at opening them. If you have a buck, he should be kept a minimum of 50 feet from does, and if possible much farther, to prevent his odor from tainting the milk. When breeding season arrives, the odor of a buck can very offensive to even the most ardent goat lover. When your mother told you that you would stink like a billy goat if you didn't take a bath, she was making the point as clear as she possibly could.

Drylots: Because fencing a large area for goats can be expensive, and because they are browsers and a field may not supply all of their food needs, some people may wish to fence a small area, called a drylot, and allow the goats to use it for exercise and supply all their food to them. If you house goats using this method, you *must* be conscientious about furnishing them with food every day. This is also an option for those with a very small space available for grazing.

Pens

Housing is simple, very much the same as you'd provide for sheep; but be more careful to avoid drafts or dampness, since goats are not as hardy as sheep. You can keep your goats with your sheep, but be on the lookout for butting activity, especially if your sheep are a horned breed, since butting may damage the udder.

Allow at least 15 square feet for each goat in your pen. As with sheep, do not have high doorsills on which a doe may injure her udder. For a larger herd, or in the case of limited space, stanchioning may be necessary. In this case allow at least an area 2½ feet by 5 feet for each goat.

Unless you have a very large herd, you won't want to go hog wild in constructing facilities, so make use of what you have for your goat pens: a part of your barn, an old shed, or a garage. The best type of flooring, for sanitary reasons and ease of cleaning, is concrete. You will, however, need a lot of bedding on this type of surface to keep the animals warm, clean, and comfortable, especially during the winter months. Wooden flooring is fine as long as you provide enough bedding to absorb the urine before it soaks into the wood. Dirt and gravel floors, with adequate bedding, are suitable and afford excellent drainage. Any of the most common and available bedding materials can be used.

Milking Room

For goat's milk that is to be marketed commercially, many regulations govern the location, size, and fittings that are necessitated. Some require an adjacent room for washing and sterilizing milking equipment, a place for cooling and storing milk, and a place for storing milking equipment. For the smaller operation, this is both unnecessary and prohibitively expensive. However, you will need to take certain steps to ensure the quality of your milk. As you probably know from experience with cow's milk in a refrigerator, milk readily collects odors from its surroundings. For this reason, you should do your milking in a room or building free from odors, and free of dust, dirt, cobwebs, and flies. You need not supply storage space for milking equipment if you store it in your house.

EQUIPMENT

You will need watering and feeding equipment, milking utensils, and perhaps a milking stand.

Hay/Grain

Goats are more particular than most animals in their eating habits. They prefer not to feed off the ground, and often won't eat hay that has fallen from the hayrack onto the ground. If you feed them alongside your sheep, the sheep might clean up after the goats. A separate grain feeder is not required because it's a good idea to feed your doe the grain ration when you're milking her. This will make her more agreeable when milking, and enable you to keep an eye on her feed intake. A hayrack of the kind used with sheep is satisfactory, as goats usually will not pull out and waste much hay with this type of feeder.

Waterers

Goats naturally need a lot of water to make milk, but they will not drink water that is even slightly dirty. Buckets or troughs should be secured to prevent tipping and be placed up off the floor a bit so droppings will not fall in them. Be sure to keep water thawed in the winter, and if the tap water is very cold, add some warmer water to it so that the goats are more likely to drink as much as they need.

Milking Equipment

The following list includes milking equipment that is both necessary and optional as indicated:

Small pail with cover (stainless steel is ideal)
Pail and cloth for washing udder
Milk strainer
Storage bottles for milk
Milk stand (optional, see below)
Scale for weighing milk (optional)

Stainless steel pails with half-moon covers are very easy to clean and easily worth the money. Properly cared for, they should last for many years. The half-moon feature is obviously designed to keep extraneous hair and other debris from contaminating the milk. Storage bottles that are also completely adequate include half-gallon fruit juice bottles or the like, provided they are thoroughly cleaned before use. While you can milk your goat on the ground while stanchioned, a milking stand, (Fig. 4.3) which greatly increases the milker's comfort, can be made of wood rather easily, Note the place for the feed dish. If space is a problem, or if you have a permanent milking room, you may want to construct a fold-away milking stand (Fig. 4.4).

FEED

While goats are among the best browsers, and will clear hardhack and other undesirable growth from pasture, don't expect them to thrive and produce milk on browse alone. They need good pasture or hay and a grain concentrate to produce large amounts of high-quality milk. The fact that goats are natural browsers makes them natural candidates for

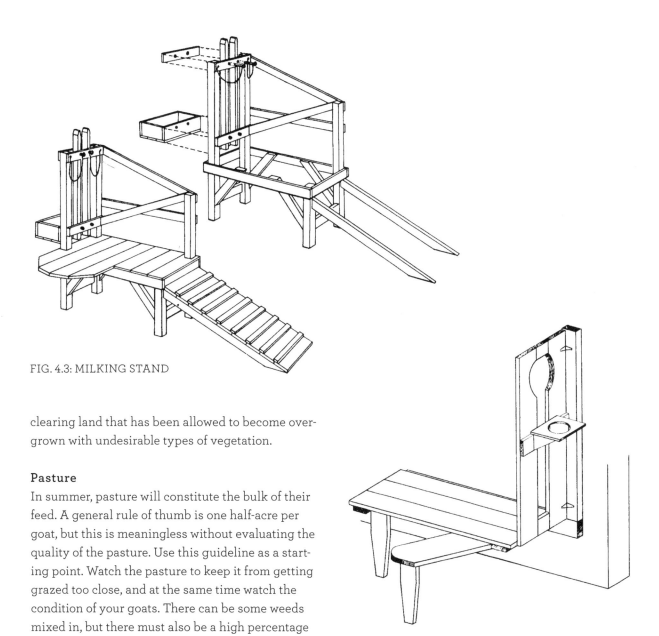

FIG. 4.3: MILKING STAND

clearing land that has been allowed to become over-grown with undesirable types of vegetation.

Pasture

In summer, pasture will constitute the bulk of their feed. A general rule of thumb is one half-acre per goat, but this is meaningless without evaluating the quality of the pasture. Use this guideline as a starting point. Watch the pasture to keep it from getting grazed too close, and at the same time watch the condition of your goats. There can be some weeds mixed in, but there must also be a high percentage of good grasses and legumes.

FIG. 4.4: FOLD-AWAY MILKING STAND

Goats do not like (nor will they thrive on) lush grasses alone. Put them on a lush, green suburban lawn and they will be miserable. If you do not have enough browse in your pasture, supply your goats with twigs, saplings, and tree prunings (especially from fruit trees) on a daily basis. Be sure you do not feed or have in the pasture any plants that are poisonous. If you raise your goats on drylot, you must supply them with as much green and browse as they will eat.

In many sections of the country, pasture is a seasonal crop and you should plan your feeding program accordingly. Pasture typically provides a high percentage of total feed intake in May and June. With the heat and low rainfall of July and August, the quality and quantity of feed available on most native-grass pastures declines sharply. With the return of cooler weather in September, good pasture is often available well into November.

During the winter, hay will constitute the bulk of your goat's diet. The better the hay, such as fine legume hay, the lower your outlay for grain will be. Your goat will need less grain and may also be fed grain with a lower protein content. A good rule of thumb is that the higher the protein content of the supplement you are feeding, the higher the cost of that feed will be on a per-pound basis.

Second-cutting hay that is at least half leguminous varieties is preferred. As with sheep, second-cutting hay is worth the extra cost per ton as there are fewer coarse stems and hence less wastage. If no legume hay is available, any good quality carbonaceous hay is suitable (see Chapter 8, "Grow Your Own"). Hay should be fed free choice, but when stocking up for winter, figure about 2 to 3 pounds per day, per goat; or about a quarter ton per goat to last for the whole season.

Dry Does (not pregnant): Dry does can be maintained on good pasture or hay, with root crops (beets, turnips, carrots) and cabbage-family plants thrown in, along with twigs from fruit trees. If any does appear in poor condition, a small amount of grain may be added to the feed program (see Table 4.1).

Milking Does: Along with the pasture and/or hay, a milking doe will need about a quarter ton of grain per year. This may be commercial grain (choose a grain that is palatable to your doe and of the proper protein percentage) or one of the mixtures listed below.

Any of these grain mixtures can be used interchangeably, and to replace commercial grain. As with all stock, do not change the feeding regimen too often or too quickly. If you must change the ration, do so gradually. The mineral mix mentioned earlier should always be available and, with it, plenty of fresh water.

For milking does, feed 1 pound of grain for each 3 pounds of milk produced (approximately 1½ quarts). One method of feeding is to keep increasing the grain slowly until peak milk production is reached, then cut the grain back (again, doing so slowly) as far as possible, without lowering milk production.

In the summer, when goats are on pasture, you need furnish the grain only in the amounts prescribed above. In the winter, supply hay free choice, plus the prescribed amount of grain mixture.

Goats will do better on a number of smaller feedings per day, with being fed a minimum of two times per day. The grain ration can conveniently be fed twice a day at milking time. This will give the doe something to keep her mind on and will make getting her into the milking stand less of a chore. Also, if she is housed with sheep, this will ensure that she gets her full grain ration and is not bullied by the sheep.

Supplementing Commercial Feed

Any of the grain mixes previously listed will lend themselves to at-home mixing using homegrown grains. In addition, silages, root crops, and cabbages can be fed to goats at a rate of 2 to 3 pounds per day, which will replace one pound of hay.

MANAGEMENT
Milking

Hold it a minute! You don't just polo the goat on the milk stand (if you use one), begin milking away, and then guzzle down the product. You must first clean and disinfect all your equipment and wash the doe's udder carefully, then milk her, strain the milk, clean your utensils, and cool the milk.

Equipment Preparation: After each milking, all milking utensils must be thoroughly cleaned. First, rinse utensils in cold water and with an alkaline detergent (do not use conventional dishwashing detergents, as they will leave residues and possibly affect the taste

Table 4.1: Grain Mixtures for Goats	
#1	Oats 60% Wheat bran 30% Soybean or linseed oil meal 10%
#2	Corn 35% Oats 35% Molasses 15% Protein supplement 15%
#3	Corn 45% Oats 20% Wheat bran 20% Soybean or cottonseed oil meal 15%
#4	Barley 50% Oats 25% Wheat bran 15% Soybean oil or cottonseed oil meal 10%
#5	Oats 100%
#6	Oats 50% Corn 50%

of the milk); place the utensils in the solution and let them soak for five minutes. Once or twice a week use an acid detergent to remove any mineral deposits that tend to build up with the use of an alkaline detergent. After soaking utensils, scrub them with a brush. You will see why seamless stainless steel buckets are easiest to keep clean. After scrubbing, rinse utensils in hot water, invert them, and allow to air dry.

Before your next milking, the utensils must be sanitized. Purchase a sanitizing chemical (chlorine, iodine, and ammonia compounds are most common—but *never* mix chlorine and ammonia, as the results can be lethal), and follow the directions that come with it. Usually utensils should be soaked in hot water containing the sanitizing agent for five minutes.

General Preparation: Before we get down to the actual milking, a few more preparations should be made. The area around the udder and rear of the goat should be trimmed periodically so long or stray hairs do not fall into and contaminate the milk. Likewise, brushing your doe before milking will remove loose hairs and dirt.

If you are feeding cabbage or silage to your does, do so immediately *after* milking so that the flavor is not passed into the milk. Feeding such feedstuffs immediately before milking may taint the milk.

Immediately prior to milking, wash the udder and your hands with lukewarm water containing a sanitizing solution; then dry thoroughly. Use paper towels for these operations since they are less likely to be contaminated than sponges or cloths you use over and over. Always handle the udder gently to prevent irritation.

If you're one of those people who have trouble tapping your fingers in sequence when you're impatient, then you may have trouble milking at first. With practice, and patience, anyone should be able to become a good milker. Be prepared, however, to annoy your doe the first few times you try it, and also to get more milk on the stand (and on yourself!) than in the bucket. Watching an experienced milker will help you learn, but if one is not available, follow these simple directions:

Be firm, but gentle, in your milking. Patting your doe before you start, and talking to her, will help you both. Grasp the top of one teat between the thumb and forefinger, as close to the udder as possible, and squeeze. This will trap the milk in the teat and prevent it from being "milked back" into the udder. Next, while pulling down gently on the teat, encircle the teat and squeeze with the middle finger, next the third finger, and finally the pinky. Done in rapid sequence, this is all there is to milking. Then do the same with the other teat, alternating teats until the udder is dry.

The first few streams of milk from each teat should be milked into a separate cup to look for "off" milk that may indicate disease.

As the milk flow lessens, pushing up on the teat and massaging the udder gently will help get the last of the milk out. After finishing milking, strip each teat by grasping the top of the teat and running the thumb and forefinger down the full length of it.

Goats should be milked twice a day (extra-heavy producers may need a third milking) as near to 12 hours apart as possible. Young does may resist milking at first, but if you are gentle and feed them grain to keep them busy and happy, they should soon settle down. A full udder is not comfortable, and the doe will soon learn that you milking will give her great relief. Above all, be consistent in your milking. Try to milk and feed at the same time each day; always milk from the same side; and if at all possible, have the same person do the milking. Treat your goat kindly and act the same way every day—leave your bad moods outside the milk room. Taking your frustrations out on your doe's udder will not be an aid to good production.

Caring for the Milk: Immediately after milking, strain the milk into a storage can or other container. Milk absorbs odors very quickly, and must be chilled to 38–40 degrees Fahrenheit as quickly as possible (ideally within 1 hour) for best quality. Chilling cannot be accomplished fast enough in a refrigerator; submerging the milk container in cold water or ice water should do the trick. Change the water frequently since it warms up easily, and when properly chilled, refrigerate the milk. A cool stream would do the job quickly and handily, and save the trouble of procuring ice cubes and changing the water.

To pasteurize or not to pasteurize? If you plan to sell your milk commercially, you will have to pasteurize. For home use or sale to consenting neighbors . . . well, it's up to you. Pasteurization, or the heating of the milk, kills disease-producing bacteria. Here is a simple, at-home pasteurization method. Small, family-sized pasteurizers are also available from farm and dairy supply outlets:

1. Place up to 6 quarts of milk in a glass or stainless steel kettle or flat-bottomed pan. (Do not use copper, iron, or chipped enamel utensils. Copper or iron utensils may cause an off flavor in the milk.)

2. Place a floating dairy thermometer in the milk. (Dairy thermometers can usually be obtained in hardware stores or dairy supply stores. Do not use candy thermometers, since these frequently have metal parts that may impart an off flavor to the milk.)

3. Heat the milk rapidly, stirring constantly with a stainless steel spoon, until a temperature of 165° F is reached.

4. Hold at 165° F for 20 seconds, then place the kettle or pan immediately into a large pan of cold water and, with constant stirring, reduce the temperature quickly to 60° F.

5. Store the milk, well covered, in clean containers in the refrigerator at 40° F.

Goats are in their prime from 4 to 6 years of age, but with proper care some does will milk satisfactorily to the age of 10 or 12.

Dehorning

Some breeds of goats are born hornless, others are not. Dehorning is a good idea, to prevent possible injury to the rest of your herd, other animals, or that favorite animal of all: yourself. Ideally, dehorning (or disbudding) should be done from 3 days to a week after birth; if much older, you may run into problems, or have to turn the job over to a veterinarian. A goat that will have horns shows a swirl of fur on its forehead; a naturally polled kid will not show any peculiarities in the hair pattern. As a further test, check the skin on the

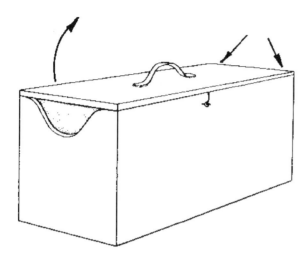

FIG. 4.5: DISBUDDING BOX

forehead. If it is loose and moves, the kid will probably not grow horns, despite even the presence of swirled hair. Tight skin on the forehead indicates a horn, and you'd better get it out soon. Nubians often exhibit signs of horns at a later age, so be careful.

The two most common methods of disbudding are using caustic paste or a hot iron. The use of caustic paste is easier, but it takes a couple of days and, if not done carefully, may run down into the eyes and cause severe damage. I do not recommend using it, but if you do, clip the hair around the buttons and surround them with petroleum jelly to protect the eyes. Secure the kid in a stanchion or in a disbudding box (Fig. 4.5) and carefully apply the paste.

After an hour, cover the paste with an adhesive bandage then let the kid go. Be sure he can't rub on other goats, and keep him out of the rain since water will make the paste run into his eyes. Watch closely to see if he rubs and spreads the paste.

Alternatively, a disbudding iron can be used to burn the tissue around each bud for 10 to 15 seconds, effectively stopping the growth of the horns.

The box shown here is invaluable in restraining a kid. It is best to have disbudding demonstrated for you by an experienced person before you try it yourself.

BREEDING

While you can get eggs from a hen without the service of a rooster, you cannot get milk from a goat without first breeding her. Does can be bred at about 7 months, when they will weigh 80 to 90 pounds. After the kidding, the kid or kids will be taken away in a few days, and the goat can be milked for about 200 days afterward. The doe should rest for 2 months and should have been bred to freshen at the end of this 2-month period, and your cycle begins again.

Bucks

In the small herd, keeping a buck is not necessary. It will cost you far more than the prevailing stud fee to keep him for the year, so there are no economic reasons for doing so.

Bucks have a characteristic odor that is, at best, unpleasant and can taint the milk. If your space and olfactory senses allow, you may consider keeping a buck for convenience, particularly if you have several does. Like sheep, does will not always exhibit signs

that are discernible to the human eye or nose when they are in heat. A number of false alarms and wasted trips to a breeding buck may convince you to buy your own buck. If you import your hired sire and have trouble discerning heat, you'll need to have him and his milk-tainting odor around for a while and expose him to the doe once a day to see if she'll accept him. This, too, can be a nuisance. In any event, try hiring a buck the first few times and see if it works.

If you decide to buy a buck, look for the qualities mentioned in selecting a milk goat. Be sure his testicles are fully descended, as well. A good grade buck can do quite well, but if you have purebred does, you should have a purebred buck of the same breed since any kids from such matings can be sold for a higher price.

Feed your buck enough to maintain his body weight, but don't let him get overly conditioned since this may make him a lazy breeder. Good pasture can maintain an inactive buck, as will hay and a pound of grain in the winter. Beginning 2 weeks before, and during the breeding season, allow 1 to 2 quarts of feed in addition to his normal ration. Provide more if heavy demands are being made of him and he looks out of condition.

While a buck is potent at about 4 months of age, don't use him for breeding until he is 7 months old. He shouldn't be used more than once every 2 weeks at that early age, but he can handle four or five does in his second year. Goats are hard to confine, and bucks that want to breed are often more difficult. Be sure to pen your buck securely away from your does when you don't want his services.

Does

Artificial insemination is becoming more widely used for goats, and you might want to check with your extension agent or local goat clubs to see if a service technician is available in your area. Like sheep, goats are seasonal breeders. They can be bred from late August to late February or as late as mid-March. They come into heat every 18 to 24 days (average: 21) for a period of 1 or 2 days during which time they'll accept the services of a buck. The gestation period is 145 to 155 days (average: 150). Table 4.2 will help you determine when your doe will kid.

While you should rest your goats (i.e., dry them off) two to three months per year, you can breed them in such a way as to never be without milk, if you have more than one doe. When they're both producing, you'll have surplus to sell or feed to other stock.

Detecting heat in does is not as easy as it may be with sows. They will exhibit much the same signs as sheep, with the additional helpful hint that their milk production may fall suddenly. Coupled with the other signs, you should be able to learn to tell the heat period with some accuracy. Remember that young bucks can impregnate does at 4 months of age, so separate them from does early on, or castrate those not wanted for breeding.

You can figure the approximate kidding time by consulting Table 4.2. If your doe is milking, you will want to dry her off about 2 months before she is due to kid. Do this by cutting the grain ration and not milking the doe. After a few days, the milk will be absorbed back into the body and she will be dried

Table 4.2: Gestation for Goats (Based on average gestation period of 150 days)		
	WILL FRESHEN: *(Breeding day less number below)**	
WHEN BRED IN:	**MONTH**	**DAY**
July	December	-3
August	January	-3
September	February	-3
October	March	-1
November	April	-1
December	May	-1
January	June	-1
February	July	0
March	August	-3
April	September	-3
May	October	-3
June	November	-3

*To determine day due to freshen, take breeding day and subtract the number indicated on the table. For example, if bred July 10 doe would be due to freshen December 7; if bred November 20 she would be due April 19.

up. Supplying plenty of roughage in the diet will aid her in this process. After she dries up, you can gradually increase her grain to a point where she is in good shape but not too fat. First-kidders that were not milking when bred need feed for their own growth as well as that of their kids. Feed to achieve the same body condition as with older does; start with 1 pound per day up to 6 weeks prior to kidding and 1½ pounds from that time to kidding.

Kidding

Kidding is almost identical to lambing, including the cutting back of bulky feeds a few days before kidding, to the actual birth process, and possible problems. The chapter on Sheep goes into considerable detail in these matters and can be used as a guideline in kidding. The major exception is that multiple births are much more common in goats. Twins after the first year are very common, and triplets and quadruplets are not uncommon.

Feeding Kids: As with all stock, make sure the young kids get the mother's colostrum. If you are going to let the mother rear the kids, then let nature take its course. However, since you have your goats for milk, it is unlikely you'll want to give up her milk for the 3 to 4 months it will take before the kids are weaned. Try not to let the kids nurse if they are to be raised artificially. First, milk some colostrum out of the mother and feed it to each kid, by bottle or eyedropper if necessary. Thereafter, milk the doe as you usually would, but don't use the milk for a few days to a week

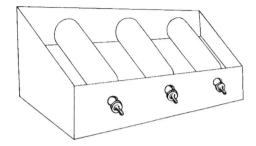

FIG. 4.6: A SELF-FEEDING BOTTLE RACK

until it reaches its normal quality (feed it to the pigs or chickens in the meantime). It will take her about a month to reach the peak of her production again.

You can raise the kids on bottles (Fig. 4.6) or from pans, but the former method is preferred since less digestive upsets from the ingestion of air will result. You can use cow's milk or a milk replacer powder (as with a veal calf) but be consistent in whatever you feed. Warm milk to 103–105 degrees Fahrenheit and feed as much as they will consume in a 10-minute feeding three to five times a day. Offer grain and good second-cutting hay at the start of the second week. Water should be provided at all times. When you wean the kids from the milk depends on their intake of hay and grain and the quality of any available pasture. It is essential that all feeding equipment be sanitized before each use to prevent disease. A rack to hold bottles can be utilized for self-feeding purposes.

For your own convenience, try to sell doe kids as soon as possible after birth. Buck kids, unless you have purebred parentage, will not be in demand for breeding and should be castrated. This can be done within the first week or two, and the procedures are identical to those for sheep. Tattooing of the ears is required for all purebred stock, and you will have to buy or borrow a tool for it if you raise purebreds. You will want to identify non-purebred kids in a litter but, rather than buying a tattooing tool, use an ear notcher or put numbered tags in their ears. In smaller litters, or in small flocks, goats can be identified by markings or by their different colors.

HEALTH

Mastitis, or inflammation of the udder, is perhaps the most serious affliction that can affect a milking goat. Symptoms may include no milk and a peculiar gait. The udder is usually quite hard and often hot, and the doe may have an elevated temperature and be partially or completely off feed.

Mastitis is usually caused by a variety of bacteria that enter the udder through the sphincter at the very end of the teat. Many of these organisms are found everywhere in the doe's environment, so the basis for control of this problem is strict sanitation. Keeping pens, yards, and pastures as clean and dry as possible will do a great deal to lower the likelihood of a doe picking up an infection when she lies down, and thus allowing the teat ends to come in contact with the ground.

Another type of organism lives in the udder and so is spread from doe to doe at milking by the milker's hands, milking machines, or any cloths used to clear the teats prior to milking. This is the contagious type of mastitis.

If a doe develops mastitis, consult your veterinarian for an appropriate treatment if you have had little experience in dealing with this condition. Often the recommendation will be to use one of several antibiotic preparations that are available for dairy cattle. These products are infused directly into the udder through the teat sphincter.

Parasite control in goats is very similar to the programs developed for sheep. Goats are vulnerable to the same types of parasites as sheep, although goats may be somewhat less susceptible to the debilitating effects of worm infestation. Developing a good deworming program for dairy goats requires very careful scheduling. Strict regulations limit the use of dewormers in dairy animals to ensure that the medication does not show up in the milk and pose a potential human health hazard. These programs involve either the use of approved medications or deworming during the dry period. When deworming is done during the dry period, the owner must be sure that the product he intends to use will not harm the developing kid.

BUTCHERING

The term chevon covers all goat meat, from the newly born kid—cutting it up almost like chicken—to the old cull doe, with the meat being used for sausage or stewing. Of course you can butcher at any stage in between. The meat is quite good and tastes like lamb, and can be cooked in the same way. Techniques of slaughter and butchering are similar to those used for sheep.

CHAPTER 5

Pigs

Pigs have long been the victims of a poor image: they have been reputed to be unclean, unfriendly, and uncouth. They are the unwilling villains of George Orwell's dystopian novel, *Animal Farm*, and their name is reserved for the most despicable of people. Happily for us, none of these rumors are true. Years ago a farmer would raise two pigs, keeping one for his family and selling the other. The income from the sold hog would pay for the other and leave extra money for "payin' off the farm." Hence the more appreciative name, "mortgage lifters."

Given the chance (that is, not being confined to a pen the size of a linen closet), pigs may be the cleanest of all farm animals. More important, they are among the best converters of feed to meat and can make use of pasturing, table scraps, and garden surpluses to reduce feed costs. In only five or six months, they will provide an average of 150 pounds of pork products. They also require relatively little space and care and are the smartest of farm animals. But beware—they are also very easy to make into pets, a decidedly negative factor if you plan to eat them. They will stand by your side for hours emitting squeals and love grunts as you scratch them. They are truly the sweethearts of the barnyard.

BREEDS

With pigs, as with most of the animals we are discussing, there is no one outstanding breed. Your choice will be based on personal preference and on what is available in your locale. Rather than trying to choose any one breed, you should concentrate on finding outstanding specimens within a given breed. While certain breeds may have their own advantages and disadvantages, any well-bred pig is suitable for home consumption.

Pigs are categorized into two broad groups: lean- or meat-type, and fat-type (Fig. 5.1). Years ago, when such "fatty" byproducts as salt pork and lard were more in demand, the "lard" pig was more common. In recent years, through selective breeding, pigs have lost the excess fat and most of them are of the leaner meat type. You should avoid the fat type. Meat-type pigs are long and trim, with hams and shoulders that are wider than their backs. These pigs will dress out with a lower percentage of fat. They are cheaper to raise, because less food goes into the production of fat than meat. They will still provide plenty of salt pork and lard for the average family.

The most common breeds of swine are:

American Landrace: The American version of the Danish hog that has made the Danes famous for their fine hams and bacon. They are white or pink with floppy ears and long, lean bodies.

Berkshire: A medium-sized hog that is very solid and has little excess fat. It is black with white on its feet and often a bit of white on its face. Berkshires have broad faces that are slightly dished out and have medium-length snouts. While this gives

FIG. 5.1: TWO MAJOR PIG GROUPINGS

them a slightly pug head, those that are extremely pug-headed should be avoided.

Chester White: A large, white hog with a medium snout and floppy ears.

Duroc: A lean, hardy hog with large, floppy ears. It is red without a trace of any other color. The color may range from deep rust to almost tan. They tend to have relatively large litters.

Hampshire: This hog is quite popular in the Northeast and is quite easy to care for; it is a good pig for beginners. It is black with erect ears and a white band around the front of the body and forelegs and, sometimes, white on the rear feet. It is a good hog for southern climates, because its darker coloration makes it less susceptible to sunburn.

Poland-China: This hog is black with white feet and splashes of white on the face and the tip of the tail.

FIG. 5.2: EXAMPLES OF COMMON BREEDS OF SWINE

It is also particularly suited to warmer climates because of its darker color.

Spot: This pig resembles the Poland-China in body type but has more white. It tends to look more like a black hog with numerous white spots.

Tamworth: One of the oldest breeds of hog, the Tamworth, like the Duroc, is red, varying from light to dark, but with erect ears and a rather long, thin snout.

Yorkshire: This is a very popular breed and also easy to keep. Yorkshires are pink or white, sometimes with black spots, and have erect ears. This is the best choice, in my opinion, if you plan to go into breeding because they have large litters and are very good mothers.

It is unlikely you will be offered, or would want to buy, a purebred pig. A purebred will cost a good deal more, and unless you're planning to go into breeding, you'll be wasting your money. I have personal experience with Hampshire-Yorkshire crosses and Yorkshire-Duroc crosses, and have found them quite acceptable. In fact, pigs, like lambs, often benefit from crossbreeding by having increased vigor and growth capability (heterosis).

PURCHASE

Pigs are usually farrowed in the spring and early fall. To start, I would suggest buying a spring shoat (weaned piglet). This pig will be raised over the summer and butchered in the fall, so there will be no need for winter housing. In addition, more table scraps and garden surplus are available in the summer, and it is easier to learn about the care of an animal during the summer. Shoats are more expensive in the spring, but it is worth the extra money to be able to raise them in the summer. It is also best to raise more than one pig at a time. If your family wants only one, see if a friend will buy another and raise it with yours. Most animals do better with companionship, and if you watch two pigs, you will notice that they never eat alone. If one gets up to eat, the other will not be far behind. In a way, they will compete for food, eat more, and tend to gain weight faster.

Ads in local papers and market bulletins offer young pigs for sale. Probably the best source of information is word of mouth. A neighbor who has had a good experience raising pigs from a nearby breeder will often be a reliable source of information about where to buy feeder pigs. These are the youngsters that have been recently weaned and are now offered for sale to anyone interested in raising one to market weight. For your first experience in raising home-grown pork, I would suggest that you go this route.

For those who fancy themselves as shrewd judges of piglets, your local auction ring will usually provide a steady supply of feeders looking for a home at which to pig out. Tracing the background on these animals is usually next to impossible. Nearby breeders often dump their excess animals at auction so they do not have to go through the expense of feeding them.

Buying pigs in this way puts the burden strictly on you, the buyer. Once the hammer falls, they are yours, and, except in very unusual cases, you have no recourse if they don't turn out as you expected.

When you buy pigs from a reputable dealer, you can ask questions about their upbringing. You might ask if any vaccinations have been given, or if the piglets have been wormed.

Many breeders who sell feeders are selling a specific breed but probably not a purebred or a specific crossbreed. In most regions of the country, the days of rather indiscriminate backlot breeders are gone. The pig business, on the commercial level, has become very specific. Crosses are developed in order that current consumer demand can be met. Today's consumer, you included, is or should be a cost-conscious individual.

Pigs properly bred and fed can deliver to the dinner table a product almost as free of fat as chicken. The national Swine Producers Association has been promoting its product as "the other white meat" and has done a good job of ensuring that its members retain their share of the national meat market.

Transport your little piglet to its new home in a small wooden box with some food inside it. Most pigs travel quite well (and do almost anything else well for that matter) if well-fed. If you have access to a dog travel crate, this will be quite satisfactory. Or you can build a simple substitute. If nothing else, they can be transported for short distances in a burlap sack. Do not make the mistake of thinking you can hold your cute little piglet in your lap on the way home. This would probably sour you on pigs forever.

Pigs are weaned at about four weeks of age and sold at 6 to 8 weeks of age. They should weigh 20–30 pounds at 6 weeks and 30–40 pounds at 8 weeks. Obviously, lighter pigs may be runts or less than 6 weeks of age, and should be avoided.

Check the going price for pigs in your area, because it is not beneath some people to sell them to you for far above the going rate if they think you'll pay it. If you can't determine the going rate, a general formula for the price of shoats is 1½ times the weight of the shoat, times the average price of dressed pork.

Piglets at 6 weeks of age have already been weaned from their mothers and are eating dry food on their own. Iron shots, which help prevent anemia, should have been administered to the piglets at birth.

Clipping the Teeth

The needle teeth of the baby pigs should be clipped soon after they erupt to avoid potential injury to the sow's teats. Piglets are very aggressive feeders, and they have no idea how rough they are while nursing. If this rough nursing takes place with sharp little teeth grasping it, the teat may become covered with minor scratches, which make the sow very apprehensive about nursing. The teeth should be cut close to the gumline with a pair of electrician's wire cutters.

There are no advantages, such as speed of growth or meat quality, in raising either a male or female pig, although some believe that males eat a lot more and consequently gain more weight.

Castration

All male pigs destined for the table should be castrated at about 6 weeks of age. A male that is allowed to grow to maturity without being castrated will have meat that has a strong odor and an objectionable flavor. Once you have sniffed it, you will never again doubt its origin.

You can tell a young barrow by one or two scars found near the hindquarters, the result of castration. If the scar is well healed, it may be necessary to feel around the scrotum to be sure that no testicles are present.

To determine the sex of the shoat, check the hindquarters. The female (gilt) has a small flap, and the male does not. In the male, the sheath that envelops the penis is readily apparent on his belly.

Look for Activity

Choose the largest and the most active pig in a litter. The belief that a pig with a curly tail is better than one with a straight tail is, in my opinion, a fallacy. Look for qualities embodied in a good meat-type hog, such as a long, lean body with large hips and shoulders.

Arrange to buy your pig early in the season, and pick it up early. Go a day earlier than suggested; the seller will sell to you then and you'll get the pick of the litter. You should refuse any pig that appears listless or sickly, coughs, or shows other signs of disease. This is also true of malformed pigs or those that are humped up, or whose vertebrae you can feel. A lump or bulge near the hindquarters often indicates a hernia, and although hernias sometimes regress on their own, choose another piglet to be sure. If you buy a barrow that has recently been castrated, make sure that there is no inflammation or any sign of infection around the incision. If buying from a stranger, be wary of any piglet that looks different from the rest. People have been known to pass a runt on from litter to litter. That alien-looking piglet may be a year-old midget.

If you know the breeder, try to observe the pigs for a few weeks before you buy, so you can choose a pig the way a breeder might choose one for himself. When I want to keep a pig from one of our litters, I look not only for one with the favorable qualities listed above, but try to select one that is especially alert, aggressive, and competitive for food. As with all animals, take your time when buying. Don't be rushed. Insist on choosing your own pig. The phrase "buying a pig in a poke" was not arrived at lightly. You can catch a piglet by quietly stalking or cornering it and quickly and decisively grabbing a hind leg and reeling it in. Ignore the shrieks and deathlike squeals—they are all just talk.

HOUSING

Pig housing can be as elaborate as you want or as your pocketbook will tolerate, but you can get by with a small area and minimal expense. Again, start simply. Unless you live in a warm climate, buy a pig in the spring for slaughter in the fall and forget about winter housing.

Three fencing materials that can be used are wood, wire mesh, and electric fencing; and three "confinement systems" are: small dirt-floored pens, small wood- or concrete-floored pens, and large pasturing pens. Depending on what land you have available and your initiative, some systems offer substantial savings in hog raising; all, however, are suitable for home hog raising.

Fencing Materials

Pigs are among the easiest animals to fence, but if they get out, especially if they are young, they can also be the hardest to catch. My preference, for ease and economy, is an electric fence. We have a small, dirt-floored wooden pen for our young weanlings until they are about 3 months of age, when we pasture them in a large, electric-fenced field. Pigs can be a problem to fence unless you are aware of the nature of the beast. If young ones escape without establishing a sense of "where the food is," you may never see them again. Older pigs rarely attempt to escape if adequately fed, and even if they do, they will not wander far from the site of their daily feeding. We have a friend who lives atop a mountain near us, and after butchering his feeder pigs in the fall, he allows his sow to range loose with his sheep.

Electric: This is my favorite because it is economical, portable, and quick to set up. You can buy an electric fence charger at your local farm supply store or through one of the numerous farm supply houses. There are two basic types: one is battery-operated and the other plugs into your electric circuit.

The battery-operated unit is best for fields a great distance from the house or barn, where running an extension cord or other electrical line would not be cost-effective.

The plug-in type delivers a constant electrical charge to the fence, and there is no worry about having to recharge the battery. When confronted with a long electrical outage, the battery-operated unit is obviously preferable.

Rather than buy metal or cedar posts, cut your own or go to the local lumber mill and buy 4-foot hardwood stakes for fencing. I use ½-inch steel (#4) reinforcing rods with plastic insulators attached by J-bolts. This is best for portable fencing, because the posts can be pulled up quite easily and the insulators can be adjusted in height to compensate for varying terrain and pigs of different ages. The metal rods can be driven into the ground with a hammer or by a simple tool constructed of two 18-inch pieces of pipe and a T-coupling (Fig. 5.3). Screw the pieces of pipe into the coupling so they form a straight handle, place the hole in the coupling over the rod, and push it into the ground.

Either metal or wooden posts can be pulled up and the whole fencing moved elsewhere to make use of available feed. Set two strands of wire, one at 6–8 inches and the other a foot off the ground. One or two inquisitive nudges with a pig's snout and the resulting jolt will discourage the most rambunctious porker. A pig is so smart that after that first run-in it may never hit the fence again, except by accident, which will only serve to reinforce the message. After a year of electric fencing, our sow needs only one wire a foot off the ground. When using electric fencing, the charger must have a good ground, and you must check regularly to be sure that weeds or rooted sod do not touch the wire and ground out the fence, thus eliminating the charge. If it does ground out, your pigs will soon discover this, and if they are not supplied with enough food they will

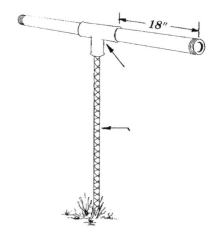

FIG. 5.3: "PIPE TOOL." *Used to drive reinforcing rod into ground.*

soon make an exit for greener pastures. All species of animals seem to have an innate sense regarding electric fences and whether or not they are operating effectively. Perhaps they are able to pick up pulsations through their hooves, which gives them a big clue as to operational status of the fence; but know, they do!

Using electric fencing for young pigs may be a problem, because the wire must be so low that it is easily grounded. For very young pigs, the bottom wire should be a bit lower than snout height and the top wire just a bit lower than the pig's height. These wire heights must be changed periodically to accommodate your rapidly growing piglets. We have had little problem fencing an entire litter of piglets in this fashion. You will find, however, that all pigs are different. Some will never go near the wire, while others will go under it all the time unless you use two strands.

Wood: If you use a wood-fenced pen, you will not be able to fence a large area as economically as with electric fencing. It is not portable and it takes more time to put up. However, it is far cheaper than any other form of fencing for a small pen. Don't be fancy—any scrap lumber will do. We had a collapsed chicken coop on our farm, and we used the sheathing boards for our first pigpen. The cost was zero.

Drive wooden posts into the ground as far apart as your boards are long, and build it to a height of 4 feet. If it's much lower than that, even a 200-pound pig could scale it if so inclined. If you house young pigs, space the boards no wider than their heads (4 inches apart or so) because if a pig's head can fit through, the rest of him will have no problem following.

Pigs will root holes several feet deep, so you must bury the bottom board a bit to prevent them from rooting out of the pen. Some people suggest ringing a pig's nose or running a strand of barbed wire along the ground to prevent rooting, but I think these are cruel practices. If burying the board doesn't do the trick, try burying some woven wire fencing a foot deep around the perimeter of the pen, or drive stakes along the outside of the pen about 6–12 inches apart.

Wire: I think that wire mesh, unless found on your farm or inherited, is only for pigs that live on New York's Upper East Side. It is so expensive that if you invest in it you will have very costly pork chops. Pen construction is the same as for wood. Make sure that the holes are small enough to contain the little fellows, and keep it close to the ground or buried 6 inches down and have it at least 4 feet high.

Confinement Systems

Pasturing: If given good pasture, a pig will graze just like a horse or sheep and may get up to half its intake this way. Using reinforcement rod posts, we make large portable pasturing pens. Shade must be provided for pigs. Either a large stand of trees or a simple lean-to or A-frame hut should be adequate. If a small stream of water is available to the pig, watering and cooling during hot weather will be simple. If not, you might have to spray your pig or sprinkle it with water on extremely hot days.

We use our pigs to rototill rough pasture, which we reseed once they have turned it over completely. If you like your pasture the way it is, make larger pens and move the pigs once the grass is eaten down, or they will begin to root extensively. Rotating pastures—that is, keeping pigs off the pasture for 2 years after grazing—should be part of any good parasite control program.

Dirt-Floored Pens: If you have a limited amount of land, hogs will do well in a smaller pen, but this will not offer you the feed economy of pasturing. Each pig should have at least 100 square feet, but allow 250 square feet or more if possible. While this doesn't afford the extra food that pasturing does, the pigs will still be able to root the ground, augmenting their diet a bit and providing themselves with much-needed iron and other valuable minerals and nutri-

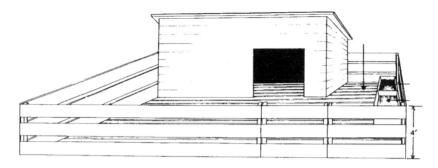

FIG. 5.4: PIGPEN WITH SLATTED WOOD FLOOR

ents from the soil. In a small pen, the ground will be rooted bare within a short time. Toss in some fresh sod or some fresh grass clippings each day.

Wood or Concrete-Floored Pens: By pouring a concrete slab or laying a floor of 2-inch boards set a foot off the ground before fencing, you will have a pen that is almost maintenance-free and escape-proof (Fig. 5.4). A quick spray with a hose every few days and your pen will be looking (and smelling) like new. This may seem to be the lazy person's pen—but don't use this system unless you wish to invest a little extra time each day in digging sod and cutting grass. If you simply feed your pig commercial feed and some kitchen scraps each day in a pen like this, I'm afraid you'll get pork not a whole lot different from what you can buy—both in taste and in price.

In the case of wood floors, slat the boards a bit to allow wastes to fall through, but do not space them so far that a pig's foot will also slip through. As with any confinement system, shade must be provided.

Winter Housing

I think of "confinement housing" as winter housing, because I would never use it except when I have to—in the winter.

Commercial hogs are often raised all year round in air-conditioned, humidity-controlled, heated houses with computer-controlled feedings. The method is very efficient, and the product will be suitable for anyone who can't tell the difference between a fresh, vine-ripened tomato and a hothouse one. I don't believe that you can raise a healthy, good-tasting hog economically using confinement housing year round. You can't discount fresh air, sunshine, and pasturing and rooting to get extra food and nutrients. Depending upon your locale, you may not need to confine your pigs even in the winter. Often a small, draft-free pig house set in your pen will be adequate. We have kept a pig through a Vermont winter with snow and 20-below-zero temperatures in such a way by supplying her with extra hay to snuggle up in. I wouldn't suggest this in such extreme climates. We did it because we didn't have time to fix space for her in the barn. She was often stiff and chilled on some below-zero mornings, and we may have been flirting with problems.

If you expect to keep a pig indoors during the winter, plan on at least 100 square feet for a sow and her litter. The shed or barn should be well ventilated to help expel the large amounts of moisture a hog emits, but it should not be drafty. A low ceiling (preferably with hay stored above) is the best for heat-conserving reasons. And the more hogs you have, the warmer it will be! Some publications suggest insulating a hog barn, but I think this is unnecessary—and expensive! As long as your pen is free from drafts and you supply your pig with extra bedding in which to snuggle, it should be fine.

Tethering: Another form of winter confinement, much like stanchioning a cow, is tethering. It is recommended only for larger numbers of pigs, but it is worth noting. You can purchase a metal tethering collar, which is secured on the pig and fastened to the ground. The pig is able to lie down and move, but is separated from its neighbor by wooden partitions. There is a food and water trough in front and a gutter for wastes behind. I don't particularly like this system because it limits the pigs, but if you have a large herd and are limited in space, I suppose there's little choice.

EQUIPMENT

You will need a feeding dish that will allow at least 1–2 feet of feeding space per pig. It should be nontippable, since pigs love to turn their dishes over, wasting food.

Old iron sinks are ideal, but while they were given away a few years ago, their price today is dependent on the price of scrap iron. There is no need to spend extra money on feeding dishes (it boosts the price of your meat, remember). Use your head. I found an old coaster wagon and a flat-bottomed barbecue grill at the landfill, and after taking their frames off, I had two fine feeding troughs. I fasten them down because they are so light, but not permanently, so that I can remove them for regular cleanings.

A number of self-feeders are designed specifically for pigs. They are well constructed to endure the relentless beating that pigs give them. They are expensive, and something less grand may have to do until you decide whether you want to stay in the pig business. There are some modifications of the store-bought types that you can probably fabricate for a very modest price. Try a 55-gallon drum. If you are clever with a cutting torch, you can cut the drum in half lengthwise and make two feeding troughs. Round over sharp edges that might cut the hungry feeders, or heat them to the point where any sharp edges are blunted.

Select a barrel that has not contained anything that might prove toxic to the pigs. Drums are often used to store materials that are used in the human food industry; such containers should be quite safe.

Old hot-water heaters can be used as troughs and prepared in the same way. They are usually rusty, but if you have the time and inclination to clean them, rust should not prove to be too big an obstacle.

Water is the cheapest of the essential nutrients

houses, roll on them, and sit on them. After you have exhausted all your old pails, buy a flexible rubber one. If you have trouble with your little friends constantly spilling their water, build a wood frame around the pail or stake it to the ground.

If you pasture your pigs far away from a water source, devise a bulk-carrying storage tank that you can fill and transport to their pen, to be stored in the shade and used to refill empty buckets.

FEED

Pigs are like humans—at least in that they have only one stomach; therefore, they are not nearly as economical as ruminant animals (goats, sheep, cows) in converting low-grade protein such as forage into meat and other high-grade proteins. Then, why are they in this book? For one thing, they are among the most efficient converters of feed to meat; second, they are about the easiest animals for which to secure *supplementary* high-quality food. Because their digestive system is similar to our own, they need a balanced diet: proteins, carbohydrates, fats, mineral salts, vitamins, and water. Table scraps, garden surpluses, and good pasture can supply all these requirements. With a little effort and careful planning, it is quite possible to raise a pig using almost no commercial feed.

Commercial Feed

Consider commercial feed, first because you will probably begin with it, and at times it will be all you have access to. Commercial pig feed is a "complete" feed, in that it contains all the food essentials listed above.

in any animal's diet. It is particularly critical for pigs and should be presented clean, cool, and in an adequate amount. Since most pigs are fed out during the warmer months when water won't freeze, a garden hose running from the house or barn will provide the water supply. Pigs quickly learn to operate self-waterers that offer a constant supply of water from a nipple-like device. These eliminate the mess of the older-style waterers. They are relatively inexpensive, have low upkeep, and are easy to install.

If you use a container, you will not want a huge one because pigs will dirty the water, which also will stagnate. A thick, pliable, 5-gallon rubber pail or a 5-gallon, wide-based galvanized tub is best, or any old bucket will do. Try to make use of what you have. The metal ones can be very hard to free from frozen water in the winter, and hard plastic ones will probably not last long, especially in colder weather, because pigs have a tendency, when their pails are empty, to bat them around, carry them into their

Table 5.1: Alternative Feed Sources for Pigs

WEIGHT OF PIG (LB)	APPROXIMATE AGE IN WEEKS	AMOUNT TO FEED DAILY
30	7	2½ lb. Grade I foods.
60	12	3½ lb. Grade I foods.
100	17	2 lb. Grade I foods plus 3 units of Grade I and II foods in any combination.
140	21	2 lb. Grade I foods plus 4½ units Grade I, II, or III foods in any combination.

GRADE I FOODS (1 UNIT = 1 LB.)

Commercial pig feed, kitchen table scraps, and swill (providing it does not contain too many potato peelings, vegetables, or too many outside leaves of green vegetables)* Restaurant and institution garbage (excellent if rich in gravies, meat, piecrust, and bread), offal, animal afterbirth (very rich in protein)**, soybeans

GRADE II FOODS (1 UNIT = 4 LB.)

Potatoes (boiled)
Sugar beets (raw)
Jerusalem artichokes (raw)
Belgian carrots
Corn ears on stalks
Most surplus garden vegetables (except as listed in Grade III below)
Young rape, young grass, clover, fresh clippings

GRADE III FOODS (1 UNIT = 10–12 LB. MAY NEED TO BOIL FOR INCREASED PALATABILITY)

Garden waste, remains of cabbage family, pea and bean vines, vegetable peelings
Cabbage, kale
Turnips, beets, beet tops
Surplus fruit

Source: Bulletin no. 18, *How to Raise a Pig*, Charlotte, VT: Garden Way Publishing Co.
Kitchen scraps and swill weigh about 2 lb. to a quart. Solid vegetables and fruits weigh about 1lb. to a quart. Greens, grasses, vines, etc. weigh about 14 lb. to a bushel.
**Fresh pork trimmings or fat must be boiled first (or frozen for three weeks) to kill any trichinae larvae.*

You can usually buy it in mash or pellet form, though pellets are often a bit more expensive. But shop around. Local feed stores often have wide differences in prices.

Most pigs like their feed moistened so it is mushy, but experiment to find out how your pig prefers his food served. Feeding mash without wetting it may make it unpalatable and may form balls of food in the pig's stomach. In the winter, I do not wet the food, for the obvious reason that it freezes solid and robs the pig of body heat. I find that feeding a pelleted ration, dry, in the winter is the best solution.

Two pigs will grow faster and more economically than one. Never, when there are two pigs around, will only one be eating. Keep an eye out for "bullying" by one pig. If one pig is getting short-changed because of the other's "hoggishness," feed them at different locations.

From weaning age to butchering weight (about 200 pounds), you will need 600–1,000 pounds of commercial feed or its equivalent (see Table 5.1). The most critical time in the raising of a pig is immediately after weaning. If a young pig does not get a good start, this will set a bad precedent for later growth. Up to 100 pounds, a pig needs feed containing at least 16 percent protein; from 100 pounds (about three months of age) to 150 pounds, it needs 14 percent protein; and from then on to butchering (200 pounds or so) it needs 12 percent protein.

Most commercial pig feeds are 16 percent protein so they will be fine for weanlings. At times I have raised the protein content even higher by supplementing their feed with a high-protein feed such as soybean or cottonseed meal. I have done this for 2 weeks after weaning to give our pigs an extra boost. By mixing the feed equally with cornmeal or whole corn after 3 months (at about 100 pounds) the protein content will be cut to about 14 percent. By increasing the corn content so that at 150 pounds it is two-thirds of the ration, you will have a feed that is about 12 percent protein.

A feeder pig (one that you raise for meat) should be fed as much food as it will clean up between feedings. If there is a lot of feed left over from the previous feeding, cut down to avoid waste. If none is left, increase until there is just a bit left at the time of the next feeding. I feed our pigs three times a day for the first 3 months and twice a day from then on. This, however, is not critical, and if three feedings are inconvenient, don't worry about it.

It is not economical to raise a pig beyond 200–225 pounds. Greedy ("piggy") people reason that more weight means more pork, but this is not the case. Gains over 200 pounds require more feed per pound of gain than below that weight, and gains are more fat than lean meat.

Unless you have a large scale, you'll have to estimate weight. This can be done fairly accurately and easily by using a weight tape or a weight formula. A weight tape, when used to measure a pig's girth, will read not in inches but in pounds. You can purchase them in some feed stores or from farm supply catalogs.

A simple and accurate formula for estimating weight is Weight = Girth x Girth x Length ÷ 400. All measurements are in inches. The girth is the measurement around the body of the pig just to the rear of the front legs. The length is from the base of the tail to a point between the ears. To make the measurements, use a tape measure or a piece of baling twine later measured against a ruler. If the weight is less 150 pounds, add 7 pounds to the total. For example, if a pig measures 40 inches in girth and 43 inches in length, it weighs 172 pounds:

$$40 \times 40 \times 43 = 68,800 \div 400 = 172 \text{ pounds}$$

Supplementing Commercial Feed

While I believe that you can raise delicious pork economically on commercial feed alone, you can realize a real savings if you supplement or eliminate the use of commercial feed by making use of pasture and the tons and tons of perfectly good and highly nutritious garbage that is thrown away to rot every day. It is probably true that what Americans throw away every day could feed half the world's starving people—any pig would attest to that. While my wife worked in a restaurant one year, our pig ate better than we did (shrimp, coq au vin, veal cordon bleu, etc.) on what the patrons discarded.

In today's environment, we need to manage the mountains of waste that our society generates daily. Much of the waste is edible, at least by a pig. Many years ago, every metropolitan area was ringed with swine garbage-feeding operations, often of considerable size. These operations utilized the garbage generated in the cities in a very efficient manner, converting it rather inexpensively into a highly desirable human food.

A series of situations and problems surfaced in the intervening years that largely eliminated these operations. But with appropriate technology and innovative management, it is conceivable that they could reappear. This idea should be explored regularly and it should give the backyard operator some management ideas.

Pasture: You can save up to 50 percent on feed by putting your pigs out on pasture. Not only is the pasture excellent food, but a pig allowed to range may correct any deficiencies in its diet. A pig will eat old trees, hardback, roots, earth, and (beware) sometimes stray chickens. It will extract valuable minerals and nutrients that will enable it to make better use of its food and help keep it healthier and disease-resistant.

You cannot expect a pig to thrive solely on pasture. Pasture is not high enough in protein and does not contain everything a pig needs for growth. You will still have to supply it with commercial mash, or other foodstuffs, as will be explained later.

A young pig can be put outside as soon as the weather is warm enough and it can be confined by your fencing. In warmer weather, a piglet can be born outside. When weaned, it will need access to high-quality, high-protein feed. An acre of good pasture will support an average of a dozen young (less than 100 pounds) pigs and six older pigs.

If you have poor pasture, as we did when we first moved to our farm, put your pigs to work. Confine them to a small area of pasture until they've rooted it bare. They will also get out all the weed and quack grass roots. In extra-rough areas, where there is hardback and small shrubs, poke a hole around the roots with a stick, and drop in some corn. The next day, you'll have a cleared spot. After it's cleared and fertilized, move them along, remove rocks, and reseed the old pasture. In time you'll have beautiful, high-quality pasture without having to pay for plowing and reseeding.

Good pasture consists of orchard grass, bluegrass, clover, and rape, or a mixture. Don't put the pigs on too soon after planting, and don't let them eat good pasture down too far or they'll begin rooting extensively. Good summer crops consist of Sudan grass, sorghum, and soybeans. Good late crops are rye, rye grass, and winter barley.

Garden Plowing: If because of hard rains, hail, or other weather-related problems, you or a neighbor cannot harvest a crop, fence it off and set your pigs out in it. This process will make good use of an otherwise wasted crop, and cut feed costs. Even after a normal harvest, a pig can be fenced in a field to glean wasted crops, stalks, and weeds. A variation of this involves the family garden. We have a strand of electric fence wire to keep out varmints and, after harvest, turn out our pigs clean up the residue. They will clean out any rotted potatoes or crops left behind as well as rooting out weeds and corn stalks. They will till the garden a foot deep and fertilize it to boot. Allowing pigs to follow corn was the standard practice in the Midwest after corn harvest,

and under the right circumstances still should be considered.

"Garbage": This is the real harvest. Perhaps I should first explain what I mean by garbage. It is not tin cans, plastic bread wrappers, or the rest of last Christmas's mince pie found in the back of your refrigerator in July. It is *fresh* kitchen scraps, plate scrapings, and scraps from restaurants or institutions, offal from butchering, stale bread or other foodstuffs that are still edible but not relished by humans, and fresh grass clippings. I understand that commercial growers cannot feed "garbage" (raw, at least), but this is our gain. The reason it is prohibited on commercial hog farms is for disease prevention, but in a small operation, with judicious management, it should not be problem. Disease, particularly trichinosis (more about this in the Health section), is caused by eating already infected pork or food contaminated with rodent droppings or the droppings from almost any fur-bearing animal. They are all trichinosis carriers. To prevent disease problems in the use of garbage, use garbage before it goes bad, and keep it in covered cans or food bins away from rodents and other vermin. Boil questionable garbage and feed only what will be cleaned up at one feeding to prevent rodent contamination.

Boiling will never be necessary if you don't collect bad stuff and if you exercise good judgment. That is, if you suddenly get access to half a ton of beautiful garbage, don't take it all. Take as much as you can to fill your containers and that you can use before it spoils. The garbage flow will never stop—

unless everyone gets with it, and into pigs. Our source of "garbage," a restaurant, is unpredictable (depending on its business), and we often freeze surplus in busy periods for use during lean times.

A wonderful program for making use of alternative feeds in raising pigs, as outlined by The Small Pig Keeper's Council of England during World War II, is given in Table 5.1. A dairy farm may be a source of milk. Dairy farmers withhold colostrum milk from the milk sold. Colostrum is the first milk that a cow gives after she calves. It is not suitable for human consumption and cannot legally be put in with other milk offered for sale. If farmers are not using this milk for feeding their own calves, pickling, or freezing it for future use, they may be willing to give it to you or sell it at a modest cost for you to feed to your pigs.

With supplemental foods, you'll have to remember that pigs are similar to humans in digestion. Feed everything in moderation. You would never eat 15 pounds of apples, nor should you feed that much to a pig. Sod fed to a confined pig is an excellent source of minerals, especially iron. For all pigs, you should supply a mineralized salt block, free choice.

Klober's *A Guide to Raising Pigs* also gives these sources of diet essentials: *Protein*: meat, fish, eggs, wheat, corn, milk; *Carbohydrates*: fruit, vegetables, cereal, milk; *Fat*: meat fats, oils; *Minerals/Salt*: soil, trace mineral salt block, sod.

"Garbage" Sources: These can be as extensive as you want. You might try planting a "pig garden." Soybeans are good feed for pigs and excellent for improving garden fertility. We threw some squash seeds into a manure pile last year and got pounds and pounds of zucchini for our pig. If you throw a whole zucchini or other hard vegetable to a pig, it probably will not eat it unless it's near starvation. However, if you puree, mash, or boil it so that it is mushy, your pig will devour it. Also, planting extra corn will pay off, since corn is one of the best feeds for finishing pork.

Most places will be happy to separate edible from nonedible refuse. As a courtesy, supply them with containers or bags, and you will never lose a source. Keep an eye out for community suppers, church suppers, and the like for sources of food. Again, supply containers and offer to help. Schools, restaurants, bakeries, food stores, and neighbors are also excellent sources.

Bakeries often charge for stale bread, but it is much cheaper than commercial feed and nearly as nutritious. A teacher friend of ours took a few minutes to draw a sign reading PIG FOOD and place it over a garbage can he placed in the school cafeteria. Some kids don't like their lunches and throw them away. More joy for your pig and your wallet.

Sort through all garbage, discarding papers, metal, and other inedibles along with bones, coffee and tea grounds, banana peels, and rhubarb tops. Pigs will often discard what is not good for them and what they don't like anyway. You will learn your pig's tastes. Your savings will be in proportion to the time you put into collecting your supplemental food. It need not be a lot—a few minutes on the way home from work can save you many, many dollars in feed costs.

MANAGEMENT

Routines

While buying spring pigs and raising them for fall slaughter is by far the most popular routine, I'd like to say a word about the alternative—late summer to winter feeding. In colder weather, a pig will only maintain weight on what would fatten a pig in warmer months. This means that more feed is required per pound of gain. For that reason this should be considered only if you have a warm pen and an ample supply of free or inexpensive garbage. On the plus side, piglets are cheaper in the fall. A good pig house for use in the coldest months will help keep your pig warmer and cut feed consumption. Plenty of fresh water is important in the colder months to ensure proper digestion.

We find such a system advantageous because we have access to plenty of fine garbage and because we eat two pigs a year, but would rather not butcher them at the same time. Many pork products, namely, sausage and cured pork products, have 4 to 6 months of recommended freezer life, so butchering twice a year gives us plenty of fresh pork.

Handling

If you've ever tried to take your pig for a walk, you'll know why this section is here. A pig has a tendency to go everywhere but where you want it to go, and often great distances in the wrong direction. For young pigs, as in picking your young shoat from the litter, corner it, grab one of the hind legs, and corral it in your arms. Holding it upside down by both hind legs often serves to quiet a shrieking piglet. As with most things, this is not guaranteed.

Larger pigs that cannot be carried or placed in burlap bags can be moved via a pig box. The box (see Fig. 5.5) measures 6 feet long, 20 inches wide, and 3 feet high and is constructed with a solid floor and slatted sides and top. The door should slide up and down in tracks formed by two ½-inch lengths of angle iron secured on each side of the door opening. Two handles extend out at either end to allow handling by four or more people. Such a box is useful for moving a sow to be bred or taking your pig to slaughter. Building the box with skids as shown will enable you to slide the box and will make movement to a truck or from pasture to pasture easier. A few morsels of tasty food will help lure a reluctant pig into the box, but I have found that after a few trips a pig will enter it of its own accord.

An old-time favorite pig-moving technique, and one that I've never had the slightest success with, is jamming a bucket over a pig's head and leading it backward. For some reason, when a pig is blinded in such a way it will usually quiet down and permit itself to be led backward by steering the tail.

One or more people with sticks can effectively move a pig by "herding." A few cracks of the stick on the side will correct any errors in the pig's navigation. Be prepared for a lot of legwork, since outrunning a pig is not as easy as it looks. Another method of moving a pig, for the stronger and more adventurous, is to tie a rope around a rear leg of the pig and, as it runs, pull the rear end around so it continues to

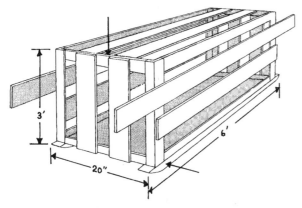

FIG. 5.5: A PIG BOX

run in the right direction. The pig supplies the loco-
motion, and you the steering.

BREEDING

I wouldn't encourage anyone to go into pig breeding
the first year in business, but it would certainly be
worthwhile considering later. Stretching the mortgage lifter concept a bit, you can keep a sow and
after a year or so pay her off by selling piglets, and
by the second year begin generating a profit that will
enable you to produce free pork for your family. In
our area, we never have problems getting rid of our
piglets and could sell many times the number we do.
Our first year we bought two gilts (females, before
they have produced their first litter), keeping the better one for breeding, and butchering the other. She
had two litters her first year, and by her spring litter
the following year, she had finished paying for herself and had begun to show a profit.

A purebred sow is not essential, but would make
your piglets that much more salable if you can get a
good deal on one. Yorkshires and Landraces are prolific and make good mothers, but almost any breed
or cross will do. A prospective sow should have a
medium-long body and a strong arched back. She
must have strong legs, feet that are well-formed and
free from injury, and she should be alert and active.
Look for a large number of teats (at least 12 or 14)
that are prominent and well spaced.

Unless you plan on a big operation or expect to
make a good income by breeding other folks' sows,
don't keep a boar. It's not like raising sheep, where
you can keep a ram for a few extra dollars a year. In
our area there are not that many boars, so we have
to settle for what's available. A boar selected for your
sow should have good conformation and have prominent testicles of equal size. Ask about his breeding
record and the size of litters. Arrange to breed with
the best boar in the area if the price and arrangement suit you (here's an opportunity to barter some
of your extra livestock). The sow will be taken to the
male. A pig box is indispensable here.

A gilt will reach puberty at five months of age,
but should weigh 200–250 pounds before she is
bred. While you can breed a sow on her first heat,
larger litters usually result when gilts are bred at
their third or fourth heat period.

Sows that are in heat may mount other sows.
The vulva becomes red and swollen. Often there is
a whitish discharge. If you push on her hams from
the top she may look lovey-eyed and stand there

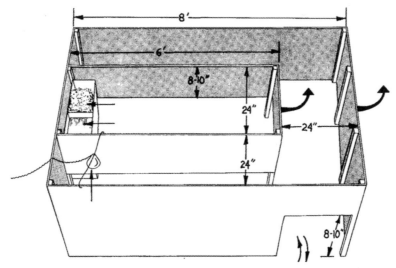

FIG. 5.6: A FARROWING CRATE

waiting. Signs vary with different pigs, and they are often hard to detect, but with experience you'll be able to determine when your sow is in heat. The heat period lasts from one to three days (the longer period for older sows) and she'll return in heat every 16–24 days until bred.

As with sheep, flushing, or increasing the feed intake prior to breeding, will usually increase the number of eggs dropped, and other things being equal, the size of the litter. While sows or gilts might normally be receiving 4 pounds of food per day, during flushing this should be increased to 5–7 pounds. Access to high-quality pasture is also recommended. You do not want a fat sow, but one that is both trim and gaining weight at the time of breeding.

The gestation period is, conveniently for those of us with poor memories, 3 months, 3 weeks, and 3 days (114 days). You can be reasonably certain she is pregnant if she doesn't return into heat. Otherwise it may be hard to see any other sign until about a week before, when she'll really begin to balloon and fill with milk. About 3 weeks before farrowing, when viewed from the side, her teats will appear enlarged and her stomach will sag a bit. When milk can be stripped from the teats, farrowing will usually occur within 24 hours. When a pig is ready to farrow, she will become very restless and begin making a nest.

If your sow has good-quality, 14–16 percent protein feed, no special supplements will be needed. While a sow should be getting about 4 pounds of good feed per day before breeding and 5–7 pounds during flushing, she should be returned to 4 pounds a day for the first three months of pregnancy. While you want to allow for the growth of the babies, you definitely don't want the sow to become fat. At two weeks to 10 days before farrowing, increase the feed intake to 1¼–2 pounds of feed per 100 pounds of pig. The larger amount should be for gilts who are not yet mature and need feed for growing as well as supporting a litter.

On the day of farrowing, some people don't feed their pigs, but I prefer to cut her feed in half. Thereafter, increase feed 1 pound per day, giving her as much as she will clean up but not exceeding 10 pounds per day. During pregnancy, exercise is also important and the sow should be fed as far from her house as possible to encourage her to move.

About a week before the sow is to farrow, you might want to prepare a farrowing crate (see Fig. 5.6). The purpose of the crate is to confine the babies near the mother immediately after birth and for the first few days afterward without danger of the sow sitting or lying on and crushing them. It is really a pen within a pen. The inner pen houses the sow and prevents her from sitting down too fast or plopping on her side. It is open 8–10 inches from the ground to allow the piglets to go in and out to nurse and to escape as the sow sits down. The other pen is fenced down to the floor and is used to confine the young pigs. On one side of the crate is a heat lamp.

I have mixed feelings about farrowing crates. I think their depends a lot on the sow—and you. If you have a gentle, calm sow and you plan to be around when she farrows, you may not need one. The first year we had a sow, I spent many hours building a perfectly wonderful farrowing crate. Our sow humored us, allowing me to keep her locked in it for a few days before she farrowed, but on the day of farrowing she wanted *out*. I let her out with visions of her crushing all her babies, but, lo and behold, she did quite well by herself. A lesson perhaps for all stock owners: They did quite all right before we were around and will probably continue to do so if we'll let them. I felt she would do better outside the crate if calm, than inside if terribly agitated.

One advantage of the farrowing crate is that you can place orphaned or rejected piglets in with a suitable sow, and she won't be able to do a great deal about rejecting them. She becomes a captive feeding station.

I think farrowing crates are best suited for larger operations, where individual attention and care are simply not possible.

If you do not have a farrowing crate, you should have a pig brooder (Fig. 5.7) with a heat lamp set in a draft-free corner. Install guard rails on the remaining sides of the pen. They should extend 8–10 inches from the wall and be 8–10 inches from the floor.

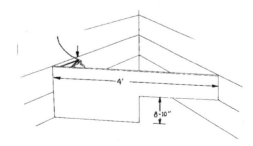

FIG. 5.7: A PIG BROODER

They permit the piglets to move about without danger of being crushed between the sow and the wall.

Once the sow appears to be near farrowing, she should be locked in the farrowing crate if you plan to use one. If it is her first litter, you should be around. Do not make unnecessary noise and certainly do not bring in the entire neighborhood to witness the event. A sow can become nervous with a lot of commotion and may eat her young or even hold back farrowing until she is alone. If you have acclimated your sow well, it will pay off many times over, because she will probably allow you to be present and help in the event problems arise.

Once labor begins, it should take about an hour or 2, but exceptions abound. (Our sow has taken as long as 6 hours.) If she has one or two piglets and none for an hour or so, or if she is in labor for a long time without having any at all, leave her alone for a while. Sows are apparently shy and hold back while people are present.

If this doesn't produce results within an hour, call a vet and explain all details. The vet may prescribe an injection of oxytocin to aid delivery, or may decide that she will need a hand (literally) in farrowing. At times a piglet will become lodged in the birth canal and a little help will be needed. Wash well before and after (before to protect the sow, and after to protect yourself) and coat your hand with petroleum jelly or KY Jelly. Bend your fingers at the knuckles so that your fingernails don't injure her, and insert your hand (you'll go pretty deep). Feel around and don't be surprised at the kicking or even nipping. You should be able to turn the infant around and gently pull it out. Either direction will do, as pigs are born both head and feet first.

I like to be present in colder weather to dry off the babies, help them get some colostrum, and place them under a heat lamp. A heat lamp should be used when the temperature drops below 50 degrees Fahrenheit. In warmer weather our pigs are born outside, and I generally leave them alone. You should remove the afterbirth, because it is not necessary for the sow to eat it and doing so might even encourage her to devour her young. If the navel cord is long and unduly hampers movement, clip to 3 inches and dip it in iodine.

A mother will usually accept her litter, and the babies will nurse about every 2 hours. If she refuses to allow them to nurse, give her a little tap to encourage her to cooperate. If she still won't have them, you'll probably lose the litter unless you want to stay awake a few nights feeding them. If you do, goat's milk is a good substitute for sow's milk. A bottle with a nipple will not be needed because piglets can be taught to drink from a pan soon after birth.

Feed the piglets every 2 hours for the first three days, six times a day for the next 4 days, and thereafter try to keep fresh milk in front of them at all times. Allow plenty of space at the feeding bowls and be sure that the bowls are heavy or the pigs will spill the milk continually. In a day or 2 after birth, you can offer them some grain moistened with milk, free choice. Water should be available at all times.

If the piglets did not get any colostrum, try to get some from the sow and feed a tablespoon to each piglet through an eyedropper. If you have another nursing sow, you might try getting her to adopt a few. (You can try this both in the case of an orphaned or rejected litter and with oversized litters in an effort to "even off" the litters.) In the case of a nervous sow who does accept her litter, enter her pen the first couple of days only for feeding, and do not organize tours with the neighborhood kids.

If you use a farrowing crate, you can let the sow and litter out of it in three days to a week, still, of course, allowing them access to the heat lamp. The light performs the double purpose of keeping the piglets warm and attracting them away from the mother after nursing. It lessens the chance of any being crushed. As the weather gets warmer and the piglets get older, the light should be extinguished periodically to wean them from it. Even if the light is not turned off, they will usually wean themselves and begin sleeping outside.

Young pigs are really quite amazing, being able to scuffle and run around an hour after birth. The first day or two are the most critical in terms of being crushed, and guard rails and the removal of heavy bedding or nest material (which the piglets might bury themselves in, out of sight of the sow, and be inadvertently crushed) will go a long way toward preventing losses.

Within 3 or 4 days, you can feed the piglets some feed mixed with warm skim milk. This is especially important with large litters to take the edge off their appetites and give the sow a break. If any runts in the litter are having difficulty getting milk, we often supplement their feed with one to three feedings of warm skim or goat's milk mixed with a little grain. We usually slip the piglet out of the pig house, run it into our kitchen, and give it the royal treatment. We have saved many a little pig this way that I'm sure would never have made it without the little extra feed we gave it.

Water given at an early age is critical for later growth. Pigs that are not given access to water early in life will learn to do without it but may require up to 25 percent more feed for growth.

Males should be castrated at 5 days of age, because the stress is less and the mother's milk is then at its peak. There is rarely more than a trickle of blood, and recovery is quick. Have on hand a new razor blade or surgical knife and a spray can of merthiolate or other antiseptic. Wash your hands well. Have someone hold the pig upside down by the rear legs with the pig's head between the holder's knees. After a few minutes of being in this position the pig will quiet down. Make an incision between the two testicles or above each one until they are exposed. Cut through the thin membrane that contains them, and push the testicles with your finger until they fall forward. When each testicle has been removed from the scrotum, grasp it firmly and twist and pull on the cord until it snaps. Watch this procedure a few times and do it under supervision before attempting it yourself. Many operations appear very simple in the hands of a skilled operator but when a novice attempts them, they become more complicated—and with less than desirable results.

Iron is essential in preventing anemia in young pigs. Iron is a component of hemoglobin, the substance that makes blood red. Part of its action is to assist in the oxygen-carrying capacity of blood. A sow's milk does not contain enough iron, and even if the piglets have access to sod they will still not absorb all the iron they need. Inject 1½ cc of iron into the fleshy part of the pig's neck at 2 to 3 days of age. Injecting in the neck is preferred over the rear leg to prevent lameness and staining the hams. Iron shots will reduce mortality, boost weaning weight, and enable the pigs to gain faster on less food.

At 4 to 6 weeks, wean the piglets from the sow. Wean the largest first and leave the smaller ones on longer to give them an added boost. You may have problems with the larger piglets accepting the late weaners, so be prepared to house them separately, or else wean them all at once. Try to wean them so that they are out of sight and earshot of the mother and vice versa. This is mostly for the piglets' sake, because after 4 or 5 weeks, the sow will be glad to get rid of them. Cut the mother's food down to normal maintenance ration at this time, and supply her with plenty of roughage to aid in drying her up. In about 21 days she will come into heat and can be rebred if so desired.

HEALTH

Swine producers were among the first groups of livestock producers to develop a sound preventative medicine program. In the 1920s, the McLean County System of swine sanitation was developed to lessen the possibility of infectious diseases being passed from one generation of pigs to the next. The basic principles of the McLean County System remain as sound today as when they were first advocated. Commercial swine operations are no longer managed in this way, but certainly for backyard or small farm operations, the principles are of considerable value.

Here is the McLean County system of swine sanitation:

1. Clean farrowing pens thoroughly and scrub them with scalding water and lye.

2. Clean the sows, particularly the udders, just before farrowing.

3. After farrowing, haul the sow and her litter to clean pasture.

4. Keep the pigs on this land until they are 4 months old.

5. Practice pasture rotation, keeping pigs off pasture for 2 successive years.

Diseases

Young pigs are particularly susceptible to roundworm (ascarids) infestations. These are worms of considerable size, and during part of their life cycle, they migrate through both the liver and the lungs. While in the lungs, they cause a condition called "thumps," in which the piglets show an exaggerated abdominal type of breathing. Pigs so affected often end up as runts.

Several good wormers are available to eliminate these parasites. Refer to the section on sheep parasites for a list of possible choices.

Several types of worms can infest pigs. Some types live in the stomach and are, quite logically, called stomach worms.

In addition to roundworms, there are several other types of worms that live in the intestinal tract, and most are effectively controlled by a combination of worming on a regular schedule and good sanitation. A simple, effective worming program can be as follows:

1. Treat sows and gilts 5 to 7 days before farrowing and again at mid-lactation (2 to 3 weeks).

2. Treat weaners before placing them on their fattening ration and then again 8 weeks later.

3. Treat fatteners at about 100 pounds.

4. Treat boars at 6-month intervals.

Vaccination

One disease that I would strongly suggest vaccinating against is swine erysipelas, or diamond skin disease. The organism that causes this disease is everywhere in the soil and can be easily prevented by a good vaccination program. The infection always seems to strike pigs at just about the time they reach butchering weight. Although pigs with the disease usually respond well to antibiotic treatment, butchering must be delayed to allow the antibiotics to clear from the meat. This means extra expense for feed, medication, and perhaps labor.

BUTCHERING

A pig 5 to 6 months old and weighing about 200–225 pounds has reached slaughter weight. Although paying someone to butcher your pig will increase the price of your pork, you should have it done at least once and watch, so you can attempt to do it yourself in the future.

Slaughtering a pig is no small operation to be taken care of between halves of a football game. The pig must be shot, stuck (bled), dipped in hot water, and have its hair scraped off. Then it is gutted, split, and hung. After the meat is thoroughly chilled, the carcass is cut up. There are several good books dealing with this process from slaughtering to curing the meat and making sausage.

Whoever does it, however, and whatever your ultimate cost, it will be the best pork you've ever tasted.

CHAPTER 6

Veal

If you are tired of paying astronomical prices for veal (or have you simply not tasted it in years?), you might be surprised to learn that you can grow 100 pounds of veal in just 3 months. No other animal we are talking about, including pigs, can give you that much meat in a single-animal operation in such a short time. While you can do it economically on commercial feed alone, it will be considerably cheaper if you have access to good cow's milk.

Veal calves require a minimum of space and time and could conceivably be raised in a suburban garage without anyone knowing—if you could explain the daily supply of rich manure you have for your garden and the occasional unusual sound emanating from within (car trouble?).

This furry little ball must be butchered at 3 months of age. Remember that by keeping his pen clean and feeding him well, you are giving him a nicer life than he would have elsewhere. Above all, do not make a pet out of him. A veal calf you decide to keep will soon eat you out of house and home.

BREEDS

While any breed can produce acceptable veal, a Holstein is preferred. A cross of a Holstein and a meat breed is also a good choice.

PURCHASE

In purchasing a veal calf, the biggest consideration is not necessarily what you buy but rather *where you buy it*. Livestock auctions are among the worst places to purchase calves. Calves at auction have been subjected to severe stress, having just been taken away from their mothers and transported

FIG. 6.1: HOLSTEIN CALF

under all extremes of temperature and usually in crowded conditions.

Another common source of day-old calves to avoid is cattle dealers who sell them off their trucks. These calves are subjected to the same stresses as those going through the auction ring, and most are candidates for disaster.

Almost all bull calves are sold soon after birth by one avenue or another. You should be able to pick one up at a neighborhood dairy farm at the going market rate. If it is possible, have the calf stay on its home farm receiving milk for about a week.

While they are more expensive, older (preferably week-old) calves are worth the extra money. If they have made it to that age, their chances of survival are considerably better. However, you still don't know how the animal was fed or handled from birth.

Your best course is to make arrangements with a local farmer to buy one of his bull calves. (Heifers are fine for veal, but dairy farmers usually

keep heifers for replacement stock, or sell them at a higher price.) Farmers often have trouble getting rid of bull calves, and you can usually get one cheaply. Make sure the calf gets colostrum from the mother for 3 days after birth, and see if it can possibly be fed the milk for a week before you pick it up. In most areas, farmers cannot market milk from cows until a week after they have calved. This milk would otherwise go to waste, but it can be fed to a veal calf. These animals will be healthier and more thrifty to feed.

Select a calf with a long, blocky body. Avoid those that are narrow and shallow-bodied and have long, crooked legs. You want alert calves that appear healthy and vigorous. Watch the calf for a while, and try to see some of its manure. If the manure is pasty and white, the calf probably has scours and should be avoided. Check the navel to be sure there is no infection there. Calves need not be castrated or dehorned, because they will be butchered before these steps become necessary.

HOUSING

Housing for a veal calf should be no problem. The chief requirement is that the calf be kept dry and free from drafts. You can probably make use of some space you already have without having to build a pen. You can use a lambing pen, a horse's tie or box stall, or a corner of a garage or barn (Fig. 6.2). Ventilation is important in the summer so that the calf does not get too hot. If the pen has windows, they can be taken out and the space covered with

Calf comfort is optimal with a calf hutch. This hutch provides year-round housing for the young calf, even in the extremes of winter weather.

burlap. This will allow air to pass through, but will keep the stall dark enough to discourage flies. The pen should be well bedded, but do not use any edible bedding such as old hay, straw, ground corn cobs, or corn stalks. Use wood chips, sawdust, or any other nonedible bedding. A pen must be thoroughly cleaned before a new calf is placed in it.

Perhaps on a visit to a dairy farm you have seen a calf hutch, a small shelter, often homemade, that is used to raise individual calves. Often these are constructed of weather-resistant plywood and are about 4 by 8 feet in size, solid on three sides, and have a door facing south, a slanted roof, and a very small wire enclosure in front to allow the calf limited outside access. These structures are ideal for raising calves under all but the most extreme weather conditions.

EQUIPMENT

Equipment is minimal. You will need a dish for feed, one for water, a salt block offered free choice, and a container for feeding milk. You can use a bottle with

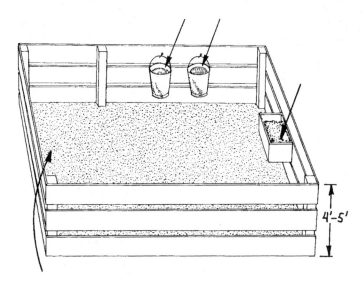

FIG. 6.2: VEAL CALF PEN

a nipple, but in time that will not hold enough milk for a feeding, so the best bet is a "nipple pail" (sold commercially under many names, among which Calf-a-teria is the most fitting). This will hold a large supply of milk. After each feeding, the milk pail and nipple should be thoroughly cleaned and sanitized with soap and hot water. Hang the bucket or bottle so the calf must raise its head—the more natural way to drink. Nipple feeding eliminates waste and also cuts down on gulping, which can lead to digestive disturbances and scours.

FEED

Some experts believe that feeding milk in the recommended quantities and allowing a grain mixture

free choice will produce a healthier, hardier animal than a milk-only diet without being detrimental to the meat itself. The meat is still pink and fries up white and tender.

Milk replacer, which when mixed with water has the consistency of fluid milk, is sold under many brand names in feed stores. Some companies market a milk replacer made especially for veal calves, but you can raise your calf on the common milk replacer almost as well. Follow the manufacturer's directions for mixing the powder. Usually it is mixed with a wire whisk with 115° F water and served at a temperature of 105°. When the calf is four weeks old, consumption is greater when the milk is served at 85°. Feed twice a day, as close to 12 hours apart as your schedule allows. If the package directions do not give adequate information for feeding veal calves (as opposed to nonveal calves, which that are being fed grain and pasture or hay), supply them with as much milk as they can finish off in 10 to 15 minutes. Do not overfeed, especially in the first few days of life (see Health section). The feeding of fresh milk will be covered in the section on Supplementing Commercial Feed.

A commercial or home-mixed grain comparable to 14–16 percent dairy fitting ration should be available free choice at all times if you choose to go the grain route. With the milk ration, the calf's grain consumption will not be very high but will be

Milk or milk replacer is essential for healthy calves and quality veal.

enough to make it a little hardier. Water and a salt block should be available at all times.

If the calf eats slowly or is off its feed, check his temperature and his manure. While the manure will be looser than that of cows or calves on conventional feed, if abnormal runniness is present, refer to Table 6.1.

If you wish to produce the ultimate product for your table, your calf will receive nothing but milk or milk byproducts. There are a variety of milk replac-ers on the market that are designed specifically for the veal calf.

The gourmet will insist that veal must be completely white, without any reddish tinge. Your tastes may not be quite as critical.

A neighboring dairy farmer may serve as a source of inexpensive feed for your vealer. Colostrum and milk from cows treated with antibiotics for a variety of reasons can be used as veal feed. In a typical situation, some combination of locally obtained and

Table 6.1: Feeding Schedule for Calves with Runny Manure

Next two feedings: withhold all feed. Replace with electrolyte solution (see Health section).
3rd and 4th feedings: normal amount of liquid, but half of normal solids. No grain.
5th feeding: normal schedule.

commercial products will turn out to be the feeding program that will be followed.

Exercise should be restricted, as the meat produced should be tender enough to be cut with a fork.

Fresh milk is naturally much better for calves, and scouring is much less common when it is fed. Feed at the temperatures recommended earlier, and feed twice a day as much as the calf will drink in 10 or 15 minutes. Stick to milk from the same farmer, since switching back and forth between herds can upset the calf's digestive system.

Calves tend to grow better at temperatures between 50 and 60 degrees Fahrenheit, so you can set your raising schedule to the times when those temperatures are likely in your region.

You may be able to realize some monetary gain in the sale of hides to a local tanner, or by selling the hide that you tan yourself. With the hair on, calfskins make fine rugs or bed coverings; without the hair, the leather is of high quality.

Handling

Handling of veal calves is not a great problem, because they rarely get larger than 200–250 pounds. A halter or collar will make handling easier and they can be made to follow by coaxing with some grain.

HEALTH

The major disease or parasite problems of veal calves are outlined for easier diagnosis and treatment in the Appendix. As with all animals, strict sanitation, correct housing, and regular feeding will go a long way toward preventing disease.

Scours is the number one problem in calf rearing, but with proper management, its effects can be minimized.

Feeding a high-quality milk replacer will do much to reduce the likelihood of scours caused or aggravated by replacers of inferior or poor quality. Infants of any species benefit from the highest-quality feed.

If diarrhea does develop, I have found it helpful to eliminate the milk or milk replacer for 24 hours and replace the calf's fluid intake with one of the numerous electrolyte solutions designed specifically for that purpose, which are available at feed stores. These preparations come in powder form and are mixed with water and fed at a rate recommended on the label or on the package insert.

The greatest hazard faced by the calf with diarrhea is dehydration. Overcoming this situation is critical if the calf is going to make it through this most serious episode. The electrolyte solutions replace the essential salts lost through diarrhea and that are necessary to good health, as well as providing glucose for energy.

Time has proved that this treatment is usually far better than antibiotics given either by mouth or by injection.

Nothing in life is simple, and so it is with calf scours. There is no easy course. Several different bacteria, viruses, and parasites may cause the calf to develop the runs, and arriving at a diagnosis is seldom easy, even for the experts. Working closely with your veterinarian, you should be able to correct the condition and develop a management scheme that will do much to avoid future problems.

Isolate the infected calves and always clean the pen thoroughly.

BUTCHERING

When the time comes, you will find that not having made friends with the little fellow will make the task much less of an ordeal. The optimal butchering age is 9–12 weeks. If your food supply will allow, you might try a few more weeks of feeding, but remember that if you allow the calf to get too old, the meat will lose the quality of veal. At eight weeks, the average weight for a Holstein calf is 160 pounds; at 12 weeks, 215 pounds. At these respective ages, Jerseys weigh 102 and 141 pounds, and Guernseys weigh 122 and 170 pounds. You can see why Holsteins are preferred. The calf should dress out to about 40 percent of live weight for the Holstein (slightly more for other breeds, since the Holsteins are bigger-boned), or about 90 pounds of veal for a 12-week-old calf.

We cannot fully describe slaughtering and butchering here. A calf, except for boning of the legs if desired, is not a difficult animal to butcher, and watching the entire process carried out by an experienced person is the best way to learn.

That little corner of your shed that once housed your lawnmower (before you traded it in for sheep) can be put to good use in the production of your veal. Like lawnmowers, expensive veal can also be a thing of the past.

Don't forget sweetbreads!

CHAPTER 7

Beef Cattle

With the ever-increasing emphasis on low cholesterol and low saturated fatty acid intake in our diets, changes are being made in the composition of the meat products reaching our dinner tables. Not too many years ago, a beef carcass that graded prime was considered the ultimate in gourmet dining. Truthfully speaking, it still is. An even fat cover and excellent marbling are still the criteria used in judging beef that will be tasty, tender, and succulent.

Nonetheless, the newer realities of our dietary knowledge have changed our buying and eating habits. Health authorities now dictate that fat consumption must be reduced to ensure our good health and to prolong our lifespans. Beef producers have developed a leaner carcass to meet these dietary recommendations. This newer beef carcass has far less waste than previous varieties and can be produced more economically than the animal that grades high on the scale as prime and choice. Better still, the small operator can raise it quite efficiently and without a major cash outlay for purchased feed.

Conventional wisdom has it that critters raised at home are almost always more tasty and more tender than animals raised in a commercial feedlot. Whether this is true is difficult to prove, because comparative taste tests are rarely attempted, and the owner/raiser is almost always biased toward the product he has carefully brought to the table.

Remember that lean beef may require a different method of cooking than meat that grades higher. The fat cover and the marbling of high-quality cuts generate the moisture that permits broiling without drying out the meat. Leaner cuts, with less fat cover, tend to dry excessively if prepared in the manner suggested for higher grades.

Generally such cuts can be cooked using moist heat, in the way one would prepare a pot roast. First, sear the cut on all sides to brown, then add water to almost cover the piece, season to taste, cover the kettle, and simmer until tender.

Smaller servings are also recommended. This is important to remember when you have the carcass custom-cut and frozen at a slaughterhouse. With smaller portions and leaner cuts, it is possible to enjoy beef on a regular basis while lessening potential health problems for you and your family.

All this brings us to the question of home-raised beef. Is it feasible, economical, and worth the effort? Like so many questions, there are no easy answers. To say that the average backyard or weekend farmer can produce it more economically than beef can be purchased in the supermarket is probably misleading, unless a number of unusual circumstances are at work in a household. But savings aren't the only consideration. If you get great personal satisfaction out of raising livestock and watching it grow and mature, a modest home beef operation should be considered.

Can You Eat It?

If you are thinking about attempting a home beef project, you have several things to consider. Is your family, both primary and extended, large enough to consume a beef carcass in 6 to 9 months? Assuming that your hypothetical steer dresses off at about

500 pounds, do you have enough mouths to eat that much meat within the limits of acceptable freezer life, say no more than 9 months? (Freezer life is that period during which the product is palatable. That doesn't mean that it cannot be eaten after that time. The meat simply is not of the same high quality as when it was frozen.)

Everyone conjures up images of sizzling sirloins or tasty tenderloins grilled over a charcoal fire when contemplating this project. Remember the realities: a lot of the meat will be hamburger and stew beef, so the ingenuity and resourcefulness of the housewife or househusband will be tested. The amount of hamburger that one can expect from a carcass varies widely, depending on the grade of the animal and the skill of the butcher, but the yield of hamburger will be close to 35–40 percent, regardless of how the butchering is done.

A common strategy is to sell or share half a carcass with a relative or neighbor. Selling gives you the opportunity to recover some of the costs of bringing the animal to an acceptable slaughter weight.

Be sure that your freezer is large enough to hold the meat when you bring it home. Half a steer is a lot of meat. It can be most disconcerting to bring home your frozen custom cuts and find that the storage space available is inadequate. Such a dilemma is sure to tax your resourcefulness. Careful planning will help you to avoid this sort of confusion. Using the rule of thumb that a cubic foot of freezer space will hold about 35 pounds, a custom-cut frozen carcass weighing 500 pounds will take about 15 cubic feet of space.

BREEDS

It may sound strange to read that we are suggesting that you purchase a dairy breed instead of a beef breed for your project. Simply stated, the primary reason is economic. The bull calf of a dairy cow is, in the majority of cases, going to be sold as soon as he is born, since he fills no real need on most dairy farms. Certainly a buyer could go to a beef operation and purchase a calf, but this approach has a couple of limitations. First, the beef operation is in the business of raising beef, so if you buy a calf, the calf's mother serves no real purpose. She is supposed to be the source of nutrients for the calf by reason of her supply of milk. This scheme then disrupts the normal flow of events in most beef operations.

A beef calf, if sold as a weanling, naturally commands a higher price than would a day-old dairy calf. In some of the smaller beef operations, the calves, once weaned, go onto full feed in order to hasten the growth process and ultimate sale for slaughter. Many people, once they get into the flow of a beef operation, do indeed seek out a favorite beef breed to raise their sights and become cow-calf operations at some level. When this stage is reached, a careful review of the several beef breeds should be made.

There are many beef breeds in this country, each of which fills a particular niche in the industry. The two breeds that probably have the highest recogni-tion are the Hereford and the Angus (Fig. 7.1). The Hereford is recognized by its white face and red coat; the Angus, by its compact body and all-black coat. These two breeds have provided the foundation stock for the cattle industry in the United States today.

The Spanish explorers arrived in North America earlier than the English, but their focus was less on colonization than on maximizing the amount of

Hereford

Angus

FIG. 7.1: EXAMPLES OF BEEF BREED STEERS

booty they could extract from the indigenous population. They did bring stock with them, which was the foundation of the cattle industry in the southwestern U.S. and California. The fabled Texas Longhorn traces its roots to the arrival of the Conquistadores.

In the post–World War II era, many so-called exotic breeds have been introduced into the beef industry to provide certain unique characteristics, including rapid growth rate of calves and finishing a carcass at an earlier age. These breeds include the Simmental and the Charolais.

The semiarid regions of the southwestern United States, with sparse grazing conditions and low annual rainfall, have brought back one of the breeds that formed the backbone of the industry in the early years: the Texas Longhorn. Its ability to survive and flourish under adverse conditions made it an early favorite. Today its lean carcass has prompted a revival in the interest in the breed, since that characteristic fits the needs of today's market.

Two breeds that are particularly suited to the subtropical areas of the country are Santa Gertrudis, developed on the renowned King Ranch in Texas, and the Brahma, imported from India. This latter breed is noted particularly for its heat resistance and tolerance to insects.

In the beef industry today, many of these breeds are crossed with one another to take advantage of the particularly strong points of each. A visit to a feedlot in the Midwest will reveal many crossbreeds that are specifically bred to grow to market weight in the shortest possible time.

For most of us starting out for the first time, the basic rule is to keep the operation as simple as possible. No fancy stuff, please. In keeping with this recommendation, I suggest that you purchase a Holstein bull calf (see Chapter 6, Fig. 6.1) for your first venture into the beef business.

I don't mean to sound prejudiced against other dairy breeds, but for beef purposes the Holstein clearly excels in terms of feed conversion. With him, you'll get the most meat for the dollar invested in feed.

The Jerseys, Guernseys, and Ayrshires do not attain the finish weight you are trying to achieve as quickly or as efficiently. Further, the so-called "yellow breeds," the Jerseys and Guernseys, possess a fat cover that is deep yellow in color. Many people find this objectionable in a hanging carcass or as a cut of beef on the dinner table.

Brown Swiss are large calves and are suitable candidates for dairy beef, but they are not as available in many areas as are the Holsteins (Fig. 7.2). Swiss tend to mature more slowly so the time required to attain suitable slaughter weight may be longer.

Why not start with a crossbreed, say a Holstein-Hereford? That's fine—if you have access to one. Many dairy farmers still breed their first-calf heifers to beef bulls, although this practice is less common today than it was even a few years ago. The reason for breeding dairy heifers to beef bulls is to try to ensure that heifers have calves of lower birth weight and consequently of smaller size, thus reducing the likelihood of calving difficulties.

PURCHASE

Dairy farmers like to get rid of their bull calves as soon as possible after birth, so most often the calves that are for sale are just a little over 1 day old. Sometimes it is possible to make arrangements with a farmer so you have the right to buy the next bull calf born on the farm. In that way, it is simply a matter of the farmer calling you when the calf is born. If you choose this approach, be sure that you have made all the arrangement to take care of the calf at home. Few dairymen want to be stuck unnecessarily with one more calf to care for.

Bull calves of the beef breeds are more available in some regions than in others, so making the decision about which type to purchase may be based on price alone.

A word of caution to those new to the business: beef steers, especially those of certain breeds, can have less desirable temperaments than the dairy breeds. Remember that I am taking about steers, not intact bulls. Mature dairy bulls are never, ever, to be trusted.

Don't buy any animal based on emotion. The runt, undersized and possibly malnourished, is an unlikely candidate for your first round in the beef business. If you are selecting from a group of calves of approximately the same age, use the same criteria you would in selecting a veal calf. Pick the largest, brightest, and most alert calf of the group. Look for strong legs, bright eyes, a glossy coat, and a naturally curious attitude.

Don't rush into a hasty decision. Take your time.

Brown Swiss

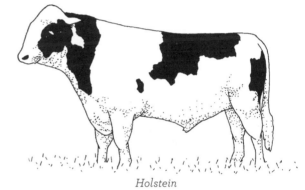
Holstein

FIG. 7.2: EXAMPLES OF DAIRY BREED STEERS SUITABLE FOR BEEF

Visit several farms and use your natural instincts to guide you.

Where shouldn't you buy beef calves? You already know the answer from reading the previous chapter on veal, but since education is a process of repetition, I will tell you again. There are two primary sources to avoid. Never buy a calf off the back of a cattle dealer's truck, and never buy one from the auction ring. Both these sources are fraught with

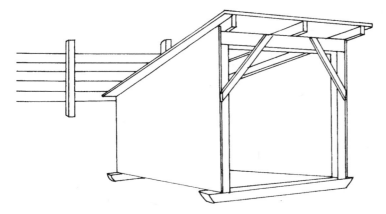

FIG. 7.3: WOODEN CALF HUTCH

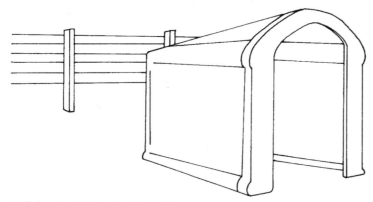

FIG. 7.4: PLASTIC CALF HUTCH

tinually using the masculine gender in this discussion, but it is an undeniable fact of life. Heifer calves of either dairy or beef background are destined for a higher calling in life, namely motherhood, while the males are generally consumed.)

Over the past several years, calf hutches have proved to be excellent shelters for dairy calves during the earliest days of their lives. They are economical, easily constructed and maintained, and ideally suited for raising the newborn to perhaps eight to ten weeks of age.

A calf hutch is a three-sided box about 8 by 4 feet, with a slanted roof and preferably open to the south (Fig. 7.3). You can modify and elaborate on this basic plan. Plastic models are available that are light, durable, easily cleaned, and free of drafts (Fig. 7.4). As is so often the case, they cost more than the homemade models.

The south-facing shelter usually opens into a small yard constructed of stock fencing or similar material that gives the calf a patio in which to air and frolic a bit. The calf is fed milk replacer, hay, grain, and water in this enclosure.

Less sophisticated than the enclosure but very adequate is a dog collar on the calf's neck and a light chain or a piece of baler twine of appropriate length securing the calf to his hutch.

potential problems and are to be avoided at all costs. Trouble is easy to find in the cattle business, so there is little point in buying it.

HOUSING

Once you have decided on your purchase, you need a place to keep him. (You must forgive me for con-

A simple box stall in the corner of your barn, coop, or shed will also do nicely, provided that its location meets a few basic needs. It should be in a bright, dry, well-ventilated, well-drained area free of drafts. If you meet these simple needs, your calf will have an adequate home.

Housing requirements for beef animals, after they pass through their first several months of life, are minimal. They can do well in a pasture without any housing. That said, I would strongly suggest that you check the animal-housing regulations in your community. Be certain that you are not violating any of them.

Be sure that pastures have an adequate number of trees to provide shade in the summer and some protection in the winter.

When no natural protection is available during the winter, especially where you are operating in a somewhat confined, modified feedlot type of operation that involves intensive feeding for fast growth, a simple run-in shed should be provided, especially where the winters are inclined to be harsh. Snow and cold don't seem to bother cattle nearly as much as wet, windy conditions. I remember very well, upon returning from the service and being hired to manage a small dairy operation, spending time trying to drive the cows into the barn on snowy nights, when in truth they were completely content lying in the snow. Having been brought up on a conventional stanchion barn operation, this new method of housing was foreign to my upbringing.

Generating body heat to offset the effects of cold, wet weather requires energy. This means that feed is consumed to create and maintain body heat. Feed used in this way is lost for use in promoting growth, so you must supply additional feed during the winter to meet this need.

FEED

Regardless of what season you start your beef, you should remember that the calf is not an efficient converter of grass into beef at an early age. In commercial cow-calf operations, the calves are on their mothers' milk until they reach weaning age. For that reason, they generally grow well early in life. Your calf will be deprived of his mother's milk, so his diet is going to be somewhat artificial, no matter what feeding program you choose.

Stick to the spring for starting your project to increase your likelihood of success. For the first weeks of life, a calf's stomach is not the efficient converter of roughage into energy that it will become later on. A newborn calf is essentially a monogastric animal. That means that he has one functional stomach that partially converts feed into a form that can be taken up by the body and utilized for growth, energy, and replacement of tissue.

You may have heard that a cow has four stomachs. More correctly stated, a cow's stomach has four compartments, in each of which certain phases of the digestive process take place.

Milk at First

The very young calf is a milk drinker. As it grows, it gains the ability to effectively utilize hay and grain for growth and maintenance. It is unlikely that any calf younger than 4 weeks effectively utilizes any feed other than milk. Somewhere after 4 weeks of age, the calf's ability to use hay and grain effectively as feed increases rapidly. It is possible to remove milk or milk replacers at this time, but it is critical to continue the grain feeding to ensure that a concentrated source of nutrients is available, remembering that under natural conditions milk would still constitute a major part of the diet.

Much of what is determined to be correct in livestock management is dependent on the expertise of the owner or caretaker, and not every situation lends itself to hard and fast rules. The Bible admonishes that the eye of the master fattens the kine (cow), and so it is in this instance. Certainly if you get the impression that your animal is not growing at a satisfactory rate, some supplementation is necessary. In any instance, 2–3 pounds of a 14 percent fitting ration a day will certainly do no harm, and the worst will be that it will cost about 30 cents a day. There is a little psychological gimmick here. If your calf knows that a little treat is coming once a day, he is apt to be much friendlier and thus easier to catch when that becomes necessary.

A 14 percent fitting ration is a grain mixture that is commercially available from most feed stores that is formulated to provide 14 percent protein; stated another way, there are 14 pounds of protein in every 100-pound bag of such mixtures. This type of grain mixture is usually composed of corn, oats, soybeans, a mineral supplement, and a vitamin concentrate. As might be expected, the ingredients contained in such mixes can vary rather widely, depending on the cost and availability of the various constituents and the manufacturer's own preferences. All states have regulations that ensure that the customer is getting the product as it is labeled.

In summary, we might suggest the following for a 500-pound steer for total daily intake, which is usually divided into morning and evening feedings:

Good-quality grass hay	6 pounds
Corn silage	15 pounds
14% fitting ration	5 pounds

For the 1,000-pound animal, the following is suggested:

Good-quality grass hay	10 pounds
Corn silage	36 pounds
14% fitting ration	10 pounds

Remember that these are suggestions and that there are endless combinations that you should explore with knowledgeable livestock owners in your area.

Some people put calves on grass too early. If you do, your beef project will no doubt get off to a poor start. Better to prolong the early hand-feeding program than to cause a potentially good beef animal to fail to achieve its maximum growth because of a poor start resulting from neglect or misinformation.

Even after a young animal is put on good grass pasture, continue to feed a grain supplement at some level. Remember your animal's ultimate purpose.

When he is a year old, your steer is better equipped to subsist on grass alone, provided that it is of good quality and sufficient quantity.

Maximum reliance on good-quality pasture is the way to economical production (see sections on pasture below). Without pasture, the costs of feeding are going to be higher, and there is no way to moderate that factor.

Measuring Feed Needs

During the late fall, winter, and early spring, when pasture is unavailable, you must supply feed to your calf to ensure its continued growth.

As an example, let's assume that your calf goes into the winter weighing about 500 pounds. How much feed will it need? A rough rule of thumb is that an animal should consume about 3 percent of its body weight daily as dry matter (DM). Dry matter is the measure that animal nutritionists use in figuring out rations for the many species with which they deal.

No feed is 100 percent DM, not even the driest of hay. Most hays are about 90 percent DM. Our 500-pound animal should be eating 3 percent of its body weight daily, or about 15 pounds of DM feed. If all that is hay, an unlikely probability, then the animal should be eating 16.6 pounds of hay daily.

Here's the formula for figuring that:

Recommended amount of
100% DM x 100 ÷ DM% of available feed
= pounds of feed required.

Here's how we use the figures in our example:

15 x 100 ÷ 90 = 16.6 pounds of feed required.

Taking this one step further, let us say that we are going to feed 5 pounds of a 14 percent fitting ration daily. If we figure 90 percent DM in the grain, then we will be supplying 5 x 0.90 = 4.5 pounds of DM as grain. This will lower the requirements for hay to 16.6 – 4.5 = 11.1 pounds. As the animal grows, we will make adjustments in the amount of feed.

Weighing Calves

How, you may ask, am I going to know how much my critter weighs to ensure that I am feeding him an adequate amount? Ideally, a scale would give you the answer. In those areas of the country where beef operations are common, portable scales are often available. These scales are an integral part of portable cattle chutes that are used to "work cattle," that is, to restrain them in order to perform necessary managerial and health procedures on large numbers of cattle. These chutes offer the restraint and flexibility necessary for range cattle that are not used to being handled on a regular basis.

If no scale is available, the next best thing is a weight tape, which can be purchased at most feed stores. Sometimes these tapes are given away as promotional materials. They are graduated in pounds. When the tape is placed around the animal just behind the front legs and snugged securely, the weight can be read directly off the tape. These tapes are accurate to within about 5 percent. For most purposes, that is more than adequate.

Having established an easy way to determine the weight of your animal, you can make the necessary adjustments in the feeding schedule.

While grass is usually the cheapest feed, another feed that is often available in livestock-producing areas is corn silage. It is an excellent source of energy, and cattle eat it readily.

To figure your animal's feed requirements, use the rule of thumb that 3 pounds of silage replaces 1 pound of hay. This being the case, you should provide 16.6 x 3 = 49.8 (make it 50) pounds of silage daily, and you should pay about one-third as much for silage as for hay, on a pound-to-pound basis.

Water

Water is the cheapest part of your feeding program. Be sure that you supply enough and that it is of good quality. Perhaps you plan to use a brook to supply the animal with its water. A roaring brook in April may be reduced to a trickle in August. Monitor your water supply carefully in the extreme heat of summer, especially if you are using a pasture with which you may not be totally familiar.

In the wintertime, be certain that the drinking trough or tub is not frozen over. This requires continual surveillance on your part. Water deprivation can have serious consequences. Almost any animal can survive for surprisingly long periods without feed, but not without water.

Several appliances are available to keep water pipes and drinking troughs from freezing. Electrical heaters and tape are the most commonly used, do not represent a big investment, and can be used for many seasons, given some minimal maintenance. The savings they represent in aggravation and time make them well worth buying. Going out to chop a hole in an ice-covered trough on a subzero morning is not everybody's idea of a good time. If you are blessed with a brook that never freezes, you are home free—but this is rarely found.

Pasture

Perhaps you are fortunate enough to have, or have access to, pasture. If you do, pasture can provide half or more of your animal's feed requirements. Grass, and grass alone, can produce the kind of carcass that meets all the requirements laid down by our cardiologists and nutritional advisers. Grass is, after all, the cow's natural diet. The ancestors of our highly refined and specialized cattle were strictly grazers. No other forage was available to them in the places where they originated.

People, in their usual protective and innovative way, modified the cow's diet to meet their own

The importance of shade for pasture grazing cannot be overemphasized. It is critical for the comfort of the animals.

unique requirements, whether the animals were meant for meat, milk, or draft purposes. Only in the very recent past, as evolutionary matters go, has the cow been provided with corn, soybean meal, cottonseed meal, oats, and a wide variety of other feedstuffs. When people tinker too much with the cow's diet, a wide range of digestive disorders can result. These disorders are rarely seen when the cow's diet is strictly grass and hay. Today, with a small but growing segment of the consuming public demanding food produced under organic methods of production, grass-fed beef fits in perfectly. The consumer can be assured that beef that is grass-fed meets the rather strict criteria laid down in organic production rules and regulations.

Pasture Growth

Native grass pasture tends to have growth patterns that relate to temperature and moisture levels, so the peaks and valleys of growth need to be monitored. In the northern reaches of the United States, grass tends to be plentiful in the late spring and early summer, then becomes somewhat dormant in midsummer, rebounding in late summer and providing good grazing in the fall, in some years well into November.

Most regions of the United States have native species of grasses that are well adapted to that particular area and its seasonal rainfall and temperature fluctuations. You must become familiar with the varieties adapted to your area in order to take advantage of the seasonal fluctuations as they occur. It is beyond the scope of this primer to identify each native species, but this in no way diminishes the importance of this understanding.

You must monitor your pasture to be certain that there is adequate feed available from mid-July to late August, and, if there isn't, to feed supplemental hay. Even a short period of water shortage coupled with excessive heat can have a most adverse effect on patterns of grass growth.

Once you have your animal gaining weight at a desirable rate on cheap feed, you don't want him drawing on his body reserves during a period of short feed.

MANAGEMENT
Avoid Stress

To ensure its best possible start in life, you should purchase your calf in the spring, ideally in April or May. Most young animals born in the wild arrive in the spring, so why break the pattern so carefully developed over the millennia?

Cattle of all ages, when subjected to changes in their environment, are stressed, regardless of the time of year these changes are attempted. When foul, wet, windy, cold weather further complicates their lives, the stress factor is multiplied, thus increasing the possibility of illness. Stress lowers any animal's resistance through a complex interaction of many factors, and this raises the animal's susceptibility to disease. Viruses and bacteria exert their harmful effects when the body's resistance is lowered through these stress factors.

Young animals generally do not have the same level of resistance to disease that their adult counterparts do, so their ability to withstand stress factors is far lower. What this all boils down to is that you should avoid moving your calf during bad weather.

There are some preventive measures that you should consider taking before moving the calf from its home farm. Ask your veterinarian what he or she would advise to help your calf make the often difficult transition from one farm to another.

We usually suggest that the calf be vaccinated prior to moving. This may lessen the likelihood of respiratory infections developing. This vaccine is administered as nose drops and is easily done in the very young calf. If you can give nose drops to a squirming 3-year-old human, you can do it to a week-old calf. The vaccine is available in single-dose units and can usually be obtained from any veterinarian who has a cattle practice. One common

trade-name product is TSV-2 by Pfizer, typically administered with a nasal cannula or a disposable plastic syringe with no needle attached.

Castration

Bull calves should be castrated early in life if they are to be raised for beef. Left intact, they will begin to "feel their oats" at about a year and a half of age. While they're seldom mean at that age, they can prove to be more than a handful. Their size makes them somewhat less than suitable playmates. Puppy dogs they are not.

Livestock scientists have conducted studies on whether castration of bulls being raised for beef improves their ability to convert feed into meat more efficiently. My conclusion is that the jury is still out, but for backyard or hobby beef, these animals should be castrated to ensure that they are as placid as possible. Having an animal castrated guarantees that his concentration will be on the task of eating as much as possible to produce the greatest possible weight gain in the shortest possible time.

There are two methods of accomplishing this: surgical and nonsurgical.

In the surgical approach, the testicles are removed. This results in little or no setback for the animal if performed when he is very young. The scrotum is cut to reveal the testicles. Each one is grasped and cut off with an instrument designed especially for this task, or the testicle is pulled from the scrotum until the cord is exposed. Then the cord can be twisted over and over, at the same time pulled until the testicle breaks free.

Obviously, no animal in his right mind is going to stand quietly while this little exercise is taking place. It is necessary to restrain the animal while this is going on. Clever operators, usually experienced livestock owners or veterinarians, can get the job done on an animal up to 2 months old by having an assistant hold the animal's tail up in the air while they do the necessary manipulations. This little trick is not for the inexperienced, or the faint of heart.

Typically the calf is cast in order to perform the operation. This means it is hogtied (the four legs are tied together) to prevent it from either running away or kicking the person performing the surgery.

This method is best learned by first assisting someone with experience, and again, done on the very young animal.

Another method is to adequately restrain the animal, then cut off the bottom third of the scrotum, break down the connective tissue between the scrotum and the testicle, and expose as much cord as possible. Then, using an instrument called an emasculator, the cord is severed. The emasculator has a special crimping edge that crushes the cut edge of the cord, thus sealing the vessels that supply blood to the testicle.

The nonsurgical technique is to use elastic bands called Elastrators, which are placed between the testicles and the body wall. These are specially made elastics, and can be purchased at most livestock supply stores. A special applicator allows you to place the Elastrator in the desired location. Over a period of several days to a few weeks, these bands restrict the flow of blood to the testicles, causing them to atrophy.

This method is easy for beginners. One word of caution: Make certain that both testicles are in the bottom of the scrotum. Not infrequently, we see situations in which one testicle is above the elastic tucked up close to the body wall and the other is in its intended location. When this happens, the novice operator is usually not aware of the problem until sometime later, at which time it is usually necessary to call in your veterinarian to finish the job. This is never a pleasant piece of corrective surgery, and you should be prepared to pay the price. Veterinarians generally do not enjoy going around behind folks who have attempted home surgery and have had their efforts fail.

Whichever method you decide to use, you should consult with your veterinarian before trying it.

HEALTH

The greatest health issue confronting the newborn calf is scours, the number one killer of calves. If you can avoid this problem by closely following the guidelines in Chapter 6 on veal calves, you have probably overcome the greatest single obstacle that you will encounter from a disease standpoint.

Most scours in very young calves are caused by a specific bacterium named *Escherichia coli,* an organism that can be found almost everywhere in nature, especially in areas where cattle are housed.

A standard practice among dairymen is to dip the navel of every newborn calf in 7 percent tincture of iodine to help prevent *E. coli* from gaining entrance into the body by way of the umbilical vein. Although this practice has been touted for years as a tremendous aid in preventing this disease, and almost all cattle raisers are aware of its importance, there are still thousands of calves born every year that suffer for lack of having this very effective, time-proven technique done. When purchasing a calf, find out if this very simple procedure has been done. There are other iodine preparations that can be substituted for straight tincture, but for the home operator the tincture still works the best.

Colostrum

You should be certain that the calf you are purchasing has received the antibody-rich colostrum from its mother as soon as possible after birth. This ensures that the animal is protected as well as possible against the large number of potential disease-causing bacteria and viruses that are always lurking in any environment. Failure to receive this critical meal often lays the groundwork for a very rocky start.

One scours-causing organism that is neither a virus nor a bacterium is called *Coccidia*, which logically enough gives rise to a disease called coccidiosis. This one-celled organism invades the intestinal tract and causes diarrhea. It can be diagnosed only by a microscopic examination of the manure. Any veterinarian can do the examination for you and prescribe a course of treatment.

Periodic fecal examinations over the life of your beefer will do much to ensure that he is never heavily parasitized. Worms, allowed to go unchecked, will do much to reduce the efficiency of feed conversion in any animal and prevent the proper rate of gain. Worming cattle today doesn't have to be an ordeal. There are a number of ways the job can be done. Perhaps the most convenient of these is the pour-on form of medication. The owner pours a measured amount of liquid, based on body weight, over the back of the animal. It is absorbed and eventually reaches the intestinal tract, where the worm population congregates. The medication goes about the business of killing the worms that are present. Ivermectin is one product often used.

Respiratory Disease

A disease that always seems to be present wherever cattle are kept is the bovine respiratory disease complex (BRDC). This complex is a group of viruses, acting singly or in combination with a bacterium that causes respiratory diseases in cattle that may range in severity from mild to acute, with death not uncommon if appropriate treatment is not given. As mentioned earlier, there are a number of very effective vaccines available that can control this complex. A check with your veterinarian will assist you in determining the best one suited to your particular area.

Respiratory diseases are most often seen in the late fall, when weather conditions become very changeable. They are also seen when the animals are subjected to any undue stress such as shipping, thus giving rise to the term "shipping fever," long associated with this syndrome.

To restate a point, I would strongly suggest not purchasing your animal when weather conditions are not the best. I would also suggest (and insist, if you were my client) that all purchased animals be vaccinated well in advance of being moved. Don't think that you can vaccinate them one day and have them safe to move the next day. Most vaccines take about two weeks to achieve maximum protection, so plan accordingly.

If you are unfortunate enough to purchase your animal and then have it come down with a respiratory condition, prompt recognition of the illness and immediate initiation of treatment are absolutely critical to a successful outcome. Left untreated, the animal faces unfavorable consequences, with the worst possible outcome being obvious and the less obvious being chronically damaged lungs, which will prevent the animal from ever realizing its full potential. Beef cattle with respiratory diseases are generally undersized, unthrifty in appearance, and have a chronic cough. When subjected to any undue exertion, they tend to fade rapidly and may collapse from the effort.

Treatment consists of antibiotics and good nursing care. The latter includes adequate shelter and access to the best-quality feed, and very minimal handling. This is difficult, for how can you medicate animals and at the same time avoid stirring them up more than is absolutely necessary? That I leave to your ingenuity and resources.

I mentioned that in many instances respiratory disease is the result of one or more viruses acting together with a bacterium. At this time, there are no commercially available medications in cattle practice that are effective against viruses, so we use antibiotics against the bacteria. Several antibiotics can be used, including the many variants on the old standby, penicillin. I strongly suggest again that you contact your veterinarian to enlist his or her advice in combating BRDC, potentially a very serious disease.

Generally speaking, the isolated backyard beef animal is not at high risk of contracting a respiratory condition, for it is usually contact with large numbers of other animals from a wide variety of different environments that gives rise to the dis-

ease. Even so, it is critical that your animal be properly vaccinated.

Pink Eye

This is an eye condition seen in cattle that occurs during the fly season. *Moraxella bovis,* the bacterium that causes pink eye, is generally spread from an infected animal to a healthy one by flies.

Early in the course of the disease, you will notice that the animal exhibits a sensitivity to light by squinting, plus an excessive flow of tears with visible staining of the hair on the nose just below the inner corner of the eye. The membranes on the edges of the eye socket become severely inflamed, and the surface of the eyeball may become cloudy or visibly whitened, depending on the severity of the condition.

The most serious threat is that the eyeball may rupture as a result of an ulcer on the surface penetrating through to the anterior chamber of the eye, with the release of its fluid content. When this occurs, the globe collapses partially or completely, resulting in blindness.

Often animals on pasture are not observed as closely as those that are confined. Furthermore, because the early stages of pink eye are rather insidious, it is possible for even the most astute, seasoned observer to miss seeing this condition. Thus it is important to check your animals on a regular basis, however cursory that observation may be.

Pink eye lends itself to treatment using a variety of antibiotics. Your veterinarian will know what gives the best results in your area. The types of preparations used include drops, ointments, and injections of the subconjunctiva. The last procedure should be left to a veterinarian.

Foot Rot

Foot rot, or infectious pododermatitis, is a disease of the hoof that, in my experience, can be seen at any season of the year, although it is generally more prevalent during wet seasons.

At least three different bacteria cause this condition, working together to cause the final result. In the typical case, the condition comes up quickly, with severe lameness and a hot, swollen hoof the most commonly observed symptoms. As you might expect, the animals are reluctant to move and do so with great difficulty.

When you find an animal showing these symptoms, examine the hoof to be certain that you are not overlooking something as obvious as a nail in the hoof or a stone or pebble lodged between the claws. With a case of true foot rot, the tissue between the claws is swollen and, as the condition advances, the tissue will become necrotic (rotten) so that it sloughs out of that space, leaving a rather deep fissure.

There are several ways to treat foot rot. The simplest is 3 to 5 days of antibiotics at a rate suitable for the size of the animal being treated. Check with your veterinarian for appropriate dosages and additional suggestions.

I like to wrap the feet of affected animals, applying a salve such as Iodex or ichthammol over the affected area. Two- or 3-inch wrapper gauze followed by waterproof tape works well. Be careful not to wrap the hoof too tightly. One way to avoid this is to use a pad of cotton as a spacer between the toes.

Hardware Disease

Hardware disease probably has been affecting cattle since the Iron Age, when people first began to discard their scrap metal in places where cattle grazed. Cattle, being possessed of great curiosity and very poor judgment, are predisposed to lap at any and all objects that capture their attention. In doing so, they are apt to swallow a wide range of materials, often to their own detriment.

The condition gets its name from the wide variety of metallic objects that have been retrieved from the second compartment, or reticulum, of the cow's stomach. When surgery is performed, it is not uncommon to recover nails, wire, staples, and sundry other objects that are potentially harmful.

The clear message is that every effort should be made to explore for such materials in every paddock, pasture, or lot where cattle are kept. When doing jobs such as roofing or fence-mending, take great care to ensure that nothing is left lying around that cattle might discover and swallow.

What typically happens is that the animal swallows a nail, which falls into the reticulum. During the normal course of things, this muscular compartment is churning away as it exerts its mixing action on the contents, which are generally the consistency of a thick gruel. It is this part of the stomach that is the tripe, often used as food. The inner surface of the reticulum resembles small squares. When a sharp metallic object is caught up in the contractions, it may penetrate the inner and outer linings of the reticulum, quickly leading to regional peritonitis, an inflammation of the lining of the abdomen. Stomach contents seep through the hole, further complicating the situation. In extreme cases, generalized peritonitis can result.

This series of events leads the animal to stop eating, since the infection causes a rise in temperature and a cessation of stomach movements, which is partially due to the pain associated with the puncture.

Typically, cattle afflicted with hardware disease develop a low-grade temperature, cease eating, assume a rather characteristic "humped-up" attitude, and are reluctant to move.

In the worst case, the penetrating object can go forward through the diaphragm and penetrate the heart—not a good situation.

Surgical treatment is often necessary. An incision is made in the rumen, and the surgeon then reaches forward into the reticulum and retrieves the nail or other penetrating object.

A more conservative, less expensive approach is to put the animal on antibiotics for several days to attempt to overcome the infection. Additionally, a magnet is given with the hope that it will find its way to the object and prevent it from penetrating any further. At the same time a laxative/rumen stimulant is often given in an effort to get the rumen contracting

again and to move out the food material that has stopped moving as a result of the peritonitis. Finally, the animal is kept as quiet as possible for several days, thus avoiding unnecessary motion and possible further penetration of the object.

Magnets are also given before they are needed to prevent objects that may be swallowed from doing their dirty work in the first place. This is a practice you should discuss with your veterinarian.

BUTCHERING

As your animal approaches 1,100 pounds, you should make arrangements to have it slaughtered. There are several ways to proceed with this final step of your beef operation.

The more adventurous of you may want to tackle the job yourself. I would caution against this unless you have a considerable reservoir of experience to draw on. Killing and dressing a steer in your backyard is a job that demands a commitment of talent and time usually beyond that of most part-time operators. Better to contract this job out to someone with the proper facilities and experience.

Most rural communities have one or more plants that do custom slaughtering, cutting, and freezing. Check with customers to determine their satisfaction with any particular slaughterhouse. By asking around, you will get a reasonable idea of who is good and who isn't.

Getting your animal to the slaughterhouse often can be arranged with your local cattle dealer, who probably has the proper equipment to get the loading done quickly and efficiently. There is no need to make the last chapter of your beef project a rodeo. The charge for this taxi service will no doubt be nominal, but whatever it is, it is probably better than something you could rig up yourself. Most cattle dealers are making daily runs to the abattoirs in your area, so it is unlikely that your request is going to put him out. Check out this method before you try loading your prize into the back of your pickup.

After slaughter, beef is typically hung in the cooler for a period of time to allow it to cure. To attain the proper level of tenderness, I recommend a minimum of 10 days for curing.

After curing, the carcass is cut up according to your specifications. Knowing the appetite and preferences of your family should allow you to suggest specific sizes of roasts, steaks, and packages of hamburger and stew meat. A good butcher will accommodate your requests exactly.

As soon as they are wrapped and labeled, the cuts are flash-frozen. The freezing completed, the plant operator will call you to pick up your meat.

I have taken you from the purchase of a calf to your freezer with customized cuts for your table. Good luck is not necessary in this operation. You must only follow the few basic principles I've discussed.

CHAPTER 8

Grow Your Own

The person assigned the pleasant task of revising this book has a different outlook on the raising of field crops by the part-time farmer than did the original author. There are, in any project, economies of time, money, and land that suggest there exist practical approaches to any situation that need careful consideration before going ahead into the great unknown. Most of us involved in the production of livestock on a more limited scale do so with definitive objectives in mind, but few of us have the luxury of unlimited time, the necessary equipment, or the expertise, to devote to our projects. Once you choose to embark on a small-scale cropping program, you must commit yourself to a great deal of hard work. This can be enjoyable, but the hard facts are that we might better channel our energies in other directions.

Given that most of us small-time livestock enthusiasts are frustrated farmers, there are additional realities we should face. On a limited number of acres, the amount of feed that can be produced will be considerably less than your livestock will need to consume. To further preserve the continuity and flavor of the original book, I have chosen to leave the author's comments and recommendations relating to the production of certain basic field crops intact, while at the same time cautioning the reader to realize the limitations these cropping programs necessarily have.

What we are concerned with are field crops and forage crops. Field crops are those plants that are primarily grown for their seeds: corn, wheat, oats, soybeans, and even sunflowers. Forage crops are the plants or parts of plants that are used for feed before maturing or developing seeds. These forages are fed as pasture (the easiest way, because the animal does all the "harvesting" itself, so there are no harvesting, curing, or storage issues to contend with), either as hay or as silage. Forage crops are further divided into legumes and grasses.

FIELD CROPS
Corn
Of the five types of corn, flint and dent corn are those that should be considered for feed purposes. Dent corn is the most widely grown feed corn, but flint corn, because it grows more quickly, is used in regions with shorter growing seasons. Plant the corn in rows 40 inches apart, with the seeds planted five to a hill and the hills 40 inches apart. Do not pick feed corn as you would a garden-variety sweet corn. Wait until after a good frost, when the husks are dry and the kernels firm. For small-scale operations, hand picking is the rule, although a husking tool will make the job faster. The stalks may be saved as they will make good bedding for some types of livestock.

We feed the whole ear (and sometimes the entire plant) to the pigs. For other animals you must husk the corn and then remove the kernels from the ear with a corn sheller. A hand sheller can be purchased inexpensively from a farm supply catalog or from your local feed store. Whole corn can be fed to livestock as indicated, or it can be cracked or made into a mash and mixed with a particular feed ration. Corn, plant and all, can also be used for silage. *Yield:* 75 to 100 bushels per acre.

Oats
Plant oats 2 inches deep and 4 inches apart, in rows that are also 4 inches apart. Oats should be harvested in the so-called "dough" stage, when they are soft but not mushy. The grain heads should be full but not so dry that they'll fall out. Cut with a scythe, preferably with a cradle attachment to catch and help pile up the oats, then rake into windrows and let dry for a day or two. Tie them in bundles and let them cure in the field for as long as it takes them to be thoroughly dry (2 days to 2 weeks, depending upon the weather). The grain must then be threshed (the grain separated from the plant) and winnowed (the chaff separated from the grain).

To thresh, place the grain 6 to 12 inches deep on a clean floor—or on a tarp, on a not-so-clean floor—and beat the plants with a flail until the grain is separated from the plants. (A flail is a wooden rod about 5 feet long with a joint made of chain about ⅔ of the way down.) After flailing, take a pitchfork and scoop out the straw, which is good to use for bedding, being sure to shake it well and release any oats still remaining in it. Winnowing is not necessary for oats that are to be fed to rabbits, as they will husk the oats themselves, but winnowing must be carried out for all other stock. To winnow, take the oats outside on a windy day and shake them from a bucket onto a tarp while standing on a stepladder. The chaff will blow away, and the heavier oats will fall down onto the tarp. Be sure it is not too windy, or you will be sowing your domestic oats wastefully and end up losing most of them. *Yield:* 30 to 50 bushels per acre.

Soybeans

Soybeans are very high in protein but are somewhat less palatable to most animals than many other livestock feeds. They are grown very much like garden beans, but require a longer growing season. Plant 6 inches apart in rows that are 3 feet apart. Allow the bean to mature fully so that it is hard. Soybeans can be fed whole if the animal will eat them, or ground up and used as a component of a feed mix.

Sunflower Seeds

Sunflower seeds are a high-protein feed that is quite easy to grow and makes excellent feed for chickens. Sow the seeds, allowing 1 foot between each plant, then let the flowers mature and go to seed. At the end of the growing season, when the seeds are fully dry, you can remove them from the flower head by rubbing it face down on ½-inch hardware cloth. The seeds can just be fed whole, as the chickens will shell them themselves.

Wheat

Get a variety of wheat earmarked specifically for livestock feed purposes. Wheat can be sown by hand, as depicted in the famous painting by Millet, *The Sower;* or by use of a seed broadcaster, which can be purchased cheaply and will spread seed quite evenly as you turn the crank and walk down the plot.

Harvesting of wheat is similar to oats, and threshing and winnowing of the seed is also necessary. *Yield:* 20 to 40 bushels per acre.

Storage

Grains can be stored in metal barrels or other covered containers to keep them free from both rodents and moisture. Make sure the seeds are dry enough before storing (10 to 15¾ percent moisture content) or they may get moldy.

FORAGE CROPS

Pasture

Pasture should be managed like any other crop. If properly cared for and of adequate size, it can provide the most time- and labor-efficient way of feeding livestock, though there are few regions of the country where pasture alone can provide year-round forage for livestock. Even in the heart of Dixie, it is difficult to get more than 10 months of grazing out of any well-managed pasture. In the northern reaches of the United States, about 6 months is a reasonably expected length of the pasture season. In the heat of summer, most pastures are of minimal value unless the acreage available is rather large and the number of animals grazing per acre is comparatively low. Midsummer droughts and excessive heat are contributing factors to poor growth of pasture grasses.

Earlier in the book it was mentioned that most grasses and legumes have preferred growing seasons. Those plant characteristics must be considered when developing a grazing plan so as not to overgraze and seriously damage the predominant plants making up the pasture. Periodic analysis of the soil is essential if one is going to return the soil those nutrients that have been taken out by the plants that have grown since the last fertilization. Your local agricultural extension agent can give you information regarding a soil test and where to send it (usually your state university). You can also visit their Web site at www.crees.udsa.gov (Cooperative State Research, Education, and Extension Service).

Do not attempt to fool Mother Nature by planting crops that are not adapted to your region. Your results will probably be poor at best. Stick to varieties that have proved to be successful in your area, and leave the experimenting to those better able to afford it. The secret, if there is one, to producing high-quality hay would be to cut it early, and elicit 100 percent cooperation from Mother Nature. For years I have looked at my mowing machine as a rainmaker, perhaps a hangover from some earlier tribal ritual. It can be the driest spring ever, but once the mower is brought out of the shed and a piece of potentially beautiful hay has been mowed, the monsoon season will follow immediately, putting a stop to all haying activities.

Harvesting

Grass, hay, and legumes should be harvested as early as possible in order to ensure the highest level of nutrients, especially protein. Suggesting this practice and achieving it are often two distinctly different things.

We have managed to be great late-June–early-July haymakers, harvesting our crop well after the peak of nutrition has passed but when the weather is more suitable for haymaking. Commercial operations try to harvest their first crop of hay as haylage (or grass silage) where they are less dependent on the vagaries of the weather. In New England, the ideal time is late May or early June, when the feed value of the crop is at its peak. Most part-time farmers have time limitations that prevent them from synchronizing their available time, often limited to weekends, with the desired amount of sunshine. The two never seem to mesh nicely, and to my knowledge the formula for doing so has yet to be determined.

As nostalgic as it might be to revert to a slower, more leisurely pace of doing this harvesting, the realities of our lives necessitate getting the haying done in as rapid and efficient manner as possible.

Buying Machinery

In most rural and semi-rural communities, a wide variety of used machinery can usually be purchased for a reasonable cost. The purchase of a small older-model tractor, simple in design, can start you on your way to doing some of the necessary work yourself. When investing in used machinery, consider your own inclinations carefully. Are you happiest when trying to

Round bales are ideal for bigger operations feeding a larger number of cattle. However, the heavy bales requires special-ized equipment to move them easily.

encourage an elderly piece of farm equipment to do its thing for one more season, or does such activity work you up into a real fit of smoldering rage? Careful now—beating on that old baler with a ball peen hammer probably isn't going to correct its malfunctioning parts.

So, if the satisfaction derived out of keeping old stuff humming along isn't exactly your cup of tea, it's probably better to hire an enterprising individual with a custom harvesting service to do the work for

you. This way you must pay to get the job done, but you are going to have to pay, one way or another, in any case. Take your choice.

Grazing

Never overgraze pastures. If you have animals that will graze very close, such as sheep and goats, have other pasture areas available so that you can rotate them from one to another and let the worn-down

FIG. 8.1: MOST ABUNDANT FORAGE GRASSES AND LEGUMES BY REGION
(SEE KEY ON FACING PAGE)

sections re-grow. Don't put your stock out to pasture too early in the spring, or the grazing will never get a good start. Check with a local feed store, nursery, or your extension agent to determine what pasture crops are most suited to your locale as well as your particular livestock needs. Good pasture (and hay) consists of a legume-grass mixture. Legumes supply protein while the grasses are high in energy. Also, the grasses will hold the soil better and prevent erosion and fill in well between the less dense legumes. Legumes also return nitrogen to the soil, while grasses and other crops deplete it.

Kale and collard greens both make good feedstuffs. These can be fed to poultry, sheep, rabbits, goats, pigs—most any livestock. They are very high in vitamins and minerals, and can even be fed to young animals, since they will not cause bloat the way many other succulent green forages might do.

Hay

Hay is any pasture grass, or legume, or a mixture of the two, cut, then dried to about a 15 percent moisture content, and stored, either baled or loose, for winter feeding (or whenever pasture is scarce or not available). A legume/grass hay mixture is best, for the reasons outlined above, but this is not to say that you must plow up all your fields and reseed them. Recent studies have shown that a well-fertilized field of non-leguminous grasses, properly cut and cured, will have nearly the same nutritional value as

Key to Fig. 8.1: Most Abundant Forage Grasses and Legumes by Region

NORTH CENTRAL AND NORTHEASTERN

Grasses	Legumes
Kentucky bluegrass	white clover
timothy	Korean lespedeza
redtop	sweet clover
orchard	alfalfa
Canada bluegrass	common lespedeza
tall oatgrass	red clover
meadow fescue	alsike clover
smooth brome	hop clover
Sudan	black medic
ryegrass	crimson clover
bents	birdsfoot trefoil
beardgrasses and bluestems	

SOUTHEASTERN

Grasses	Legumes
Bermuda	common lespedeza
carpet	hop clover
dallis	white clover
vasey	Persian clover
redtop	black medic
Bahia	spotted burclover
fescue	
Rhodes	

GREAT PLAINS

Grasses	Legumes
grama	sweet clover
wheatgrass	alfalfa
buffalo	
brome	
galleta, tobosa, curly mesquite	
bluegrass, bluestem	
Sudan	
bluegrass	
timothy	
redtop	
fescue	
Rhodes	

INTERMOUNTAIN

Grasses	Legumes
wheatgrass	alfalfa
grama	white clover
brome	sweet clover
galleta, tobosa, curly mesquite	alsike clover
red clover	
black medic	

SOUTH PACIFIC COAST

Grasses	Legumes
fescue	California burclover
brome	alfalfa
wild oats	white clovers
Bermuda	black medic
Sudan	strawberry clover
bluegrass	
timothy	
redtop	
beardgrasses and bluestems	

NORTH PACIFIC COAST

Grasses	Legumes
ryegrass	red clover
Kentucky bluegrass	white clover
bent	hop clover
red canary	alsike clover
orchard	
meadow fescue	
redtop	
timothy	
tall oatgrass	
meadow foxtail	

a legume/grass hay blend. This again may make it cheaper in the long run, because legumes generally have to be reseeded more often than grasses.

Second-cutting hays (because they are more tender and have less hard stems) make the best feed and less material is wasted. Most of the nutrients are concentrated in the leafy portions of the plants, so hay that is cut when the plants are most leafy, especially those from well-fertilized fields, offer the most nutrition. Table 8.1 illustrates the importance of an early cut and proper curing. It can be used as a guide, both when making your own hay or when choosing hay that is offered for sale.

Silage

Silage is crops such as grasses, legumes, corn (whole plants, as well as the ears) and the like, chopped, stored, and fermented in the absence of air. After the silage is packed into the silo or other airtight container, the oxygen is used up in a short time and anaerobic bacteria produce lactic and other acids. As the acids increase, the bacteria die off; as long as no more air is permitted to enter, the process halts and the silage will keep almost indefinitely.

If you choose to incorporate it into your feeding program, silage can often be purchased at a reasonable price from neighboring farmers who may have a surplus for sale. On a very small scale, the aggravation of putting up silage yourself is hardly worth it. Of course if you want to give it a try, it can be done.

While the large silos attached to barns are most common for making silage, these are much too large

Table 8.1: How to Determine Hay Quality—Cornell Study			
#1	*Stage of growth when cut* (the more mature, the more loss of nutritive value)		
		Digestible Protein	*Energy as T.D.N.*
	a. vegetative stage	18.7%	70%
	b. bud stage	14.5%	63%
	c. bloom stage	10.2%	56%
	d. mature stage	6.4%	49%
#2	*Leafiness*—important (more proteins and vitamins in leaves)		
#3	*Green color*—very important a. need 35%–60% of original color b. lose color—lose 90% vitamin A c. one or two rains on dry hay causes loss of 40%–60% feed value		
#4	Compare alfalfa, a legume hay, and timothy, a grass hay, as to feeding value. Notice that the later grasses are cut, the lower the protein content.		
		Digestible Protein	*Energy as T.D.N.*
	a. early bloom timothy	4.2%	51.6%
	b. full bloom timothy	3.2%	48.0%
	c. late bloom timothy	2.4%	47.5%
	a. leafy alfalfa	12.1%	51.1%
	b. good alfalfa	10.3%	51.1%
	3. stemmy alfalfa	8.2%	47.5%

Source: *Dairy Goats, Breeding/Feeding/Management.* New England Dairy Goat Industry leaflet 439, Amherst: Univeristy of Massachusetts.

for the small-scale type of operation we are talking about. Since the most important consideration in making silage is the absence of air, any reasonably airtight container can be used. Fifty-gallon drums are a good size and the most readily available type of container.

Since silage will begin to spoil as you use it, utilizing a number of drums will enable you to use one container at a time and not risk spoiling all your crop. Of course, any other airtight container, smaller or larger, can be employed to suit your needs. If you really get into making silage and find it affords you sufficient savings on feed costs, then you might want to construct a larger, more permanent silo. A large circular silo (square containers are used less often because of the difficulty of eliminating all the trapped air from right-angled corners) made out of corrugated steel roofing fitted with a tight top and set on a concrete slab can be made cheaply and easily. If you place an airtight door at the bottom, you can remove the silage with a shovel without worry of spoilage.

Whether you use drums or other silos, the procedure for making silage is the same. The moisture content of the material to be ensiled is very critical. It should be between 55 and 65 percent moisture (as a guide, freshly cut hay is about 75 percent moisture); too moist and it will spoil easily; too dry and it will be unpalatable to your livestock. In order to pack and ferment well, the material should be chopped. Farmers use a silage chopper, but you will have to do this by hand or use a shredder or composter if you have one. It is best to have it cut into ½-inch lengths, or be finely chopped. Pack it tightly into your silo and then be sure to fit it with a tight cover. Every day for a week, stomp on it and press it down to exclude any trapped air. Do this until the mixture is settled. Be sure the top is on tight and you should have a minimum of spoilage. Some people add molasses as a preservative but this is not essential. Your silage should be ready in about 3 weeks.

You will always have a little spoiling where air has leaked in after you begin removing the silage. Be sure to remove any moldy or spoiled silage before feeding it, and check your batch each time you remove some to be sure there isn't any undue spoilage.

Seeds and Fertilizing

Buy your seeds locally; these will be most adapted to your particular growing region. Check with local feed stores and, more important, your local extension agent, to determine what crops and varieties are best suited for your region.

Have your soil tested before sowing your crops, and then periodically every few years after to determine what is needed to maintain soil fertility, and the optimum growth and health of your pasture. Most state universities test soil samples for a nominal fee. They will recommend the correct amounts of chemical or organic fertilizers, depending on which you state as your preference.

Table of Useful Facts

ANIMAL	FEMALE	MALE	YOUNG ANIMAL	CASTRATED MALE	NORMAL TEMP. °F
Chicken	Hen	Rooster	Chick[1]	Capon[2]/Stag[3]	107.5° (young: 102–106°)
Duck	Duck	Drake	Duckling	n.a.	
Goose	Goose	Gander	Gosling	n.a.	
Turkey	Hen	Tom	Poult	n.a.	
Sheep	Ewe	Ram/ Buck	Lamb	Wether	100.9–103.8°
Goat	Doe	Buck	Kid	Wether	101–102°
Pig	Gilt[6]/Sow[7]	Boar	Shoat/Piglet	Barrow[2]/Stag[3]	101.6°/103.6°
Rabbit	Doe	Buck	Kindle	n.a.	102.5°
Cow	Heifer/Calf	Bull	Calf	Steer	98–102.5°

ANIMAL	AGE OF PUBERTY	INCIDENCE OF HEAT	DURATION OF HEAT	GESTATION	AVERAGE PRODUCTIVE LIFE*
Chicken	4–6 mos.	n.a.	n.a.	21 days	2+ yrs.
Duck	5–7 mos.	n.a.	n.a.	28 days[4]	3 + yrs.
Goose	5–7 mos.	n.a.	n.a.	29–31 days[5]	3+yrs.
Turkey	n.a.	n.a.	n.a.	28 days	3 + yrs.
Sheep	9 mos.† (avg. 16½)	13–19 days	3–72 hrs. (avg. 147)	144–155 days	10–12 yrs.
Goat	7 mos. (avg. 21)	18–24 days	1–2 days (avg. 150)	145–155 days	10–12 yrs.
Pig	3–5 mos. (avg. 21)	16–24 days	1–3 days	114 days	8–9 yrs.
Rabbit	6–7 mos.[8]	See text	See text (avg. 31)	29–35 days	3–4 yrs.
Calf	n.a.	n.a.	n.a.	n.a.	n.a.
Beef heifer	8 mos.	21 days	18–21 hrs.	283 days (+/- 3 days)	10–14 yrs.

*This is an average period of good production. This is not to say a ewe will not give birth to a live lamb after the age of 12, or a goat will definitely not milk after the age of 12; rather, these are guidelines to help you determine how many productive years an animal will give you on the average. In the case of poultry, while they will lay after the age of 2 or 3 years, their egg production will have decreased so markedly as to make them uneconomical, even for the backyard farmer.

† See text
[1] Female: pullet; Male: cockerel
[2] Castrated before sexual maturity
[3] Castrated after sexual maturity
[4] Muscovy, 35 days
[5] Canadian and Egyptian, 35 days
[6] Before having a litter
[7] After having a litter
[8] Medium-weight rabbit breeds

Disease Tables

It is not the purpose of this summary to list every disease of each of the species described in this book, but rather to give the reader a brief overview of the more common diseases that affect the species discussed.

POULTRY

INFECTIOUS DISEASE	POULTRY AFFECTED	CAUSE	SYMPTOMS	TREATMENT	PREVENTION
Brooder pneumonia	Young chicken and turkeys especially	Fungus: *Aspergillus fumigatus*	Rapid breathing, labored, depression	None	Strict sanitation in brooder equipment
Blackhead or histommoniasis	Turkeys especially	Protozoa: *Histoma meleagridis*	Listless, ruffled feathers, dark blue head	Carbasone, nitarsone, dimetridazole, ipronidazole, acidified copper sulfate, plus others	Strict sanitation; rotate ground; keep other birds away
Botulism Lembunich	Ducks	Bacteria: *Clostridium botulinum* (toxin)	Sudden death	None	Sanitation
Coccidiosis	All species	*Coccidia,* one or more variety for each species of bird	Bloody droppings in chickens	Amprolium, sulfonamides; get the current recommendations	Sanitation
Duck plague	Chicks, geese	Herpes virus	Sudden death		Keep wild fowl away from flock, and vaccination

INFECTIOUS DISEASE	POULTRY AFFECTED	CAUSE	SYMPTOMS	TREATMENT	PREVENTION
Fowl cholera	All	Bacteria: *Pasteurella multocida*	Rapid death	Sulfaquinoxaline, tetracycline	Vaccination
Fowl pox	Chickens, turkeys	Virus	Poxlike sores on unfeathered areas of body	None	Vaccination, in areas where disease is a problem
Viral hepatitis of ducks (DVH)	Ducks under 7 weeks	Picornavirus	Sudden death	None	Rat and wild duck control; strict isolation of young ducks; vaccination of adult breeder stock from older birds
Hemorrhagic enteritis of turkeys	Young turkeys	Group II adenovirus	Sudden onset, bloody diarrhea	None	Vaccination (given in drinking water) enteritis of turkeys
Infectious bronchitis	Chickens	Coronavirus	Coughing, tearing	None	Good nursing, vaccination
Infectious coryza	Chickens	Bacterium: *Hemophilus gallinarum*	Facial swelling may close eyes, discharge	Erythromycin	Buy replacement from clean flock; a bacterin is available
Infectious laryngotracheitis	Chickens	Virus	Gasping, coughing with hock extended	None	Vaccination
Influenza	All	Virus	Signs are those of a cold	None	Reportable disease
(Note: Avian flu—especially the H5N1 strain, which has become such a concern lately—falls into this category, but so far is limited to Asia, Europe, Africa, and the Middle East.)					
Infectious serositis, "new duck disease"	Primarily ducks	Bacteria: *Pasteurella antipestifer*	Eye and nose discharge	Sulfaquinoxaline, penicillin, steratomycin	Bacterin available
Newcastle disease	All	Virus	Coughing, sneezing, with nervous system involved in severe cases	None	Vaccination

INFECTIOUS DISEASE	POULTRY AFFECTED	CAUSE	SYMPTOMS	TREATMENT	PREVENTION
Pullorum disease	Chickens, turkeys, ducks	Bacteria: *Salmonella pullorum*	Symptoms in young poultry: listless, whitish diarrhea, do not eat	None	Control by testing of breeding flocks; eradication program; vaccination
Marek's disease	Chickens	Virus. One of the first cancers proven to be caused by a virus	Paralysis, sometimes only depression	None	Vaccination

RABBITS

INFECTIOUS DISEASE	CAUSE	SYMPTOMS	TREATMENT	PREVENTION
VIRAL DISEASES				
Infectious myxomatosis	Virus; poxvirus	Milky eye discharge; ear edema; nasal discharge	None	None
BACTERIAL DISEASES				
Pasteurellosis, "Snuffles"	*Pasteurella multocida*	Thin, purulent nose and eye discharge	None	Recovered animals may become carriers
Abscesses		Abscess on any part of body	None	
Mastitis (Blue breasts)	*Staphylococci, Streptococci*	Hot, swollen mammary glands	Penicillin variants	Good sanitation
Treponematosis, vent disease	Spirochete: *Treponema cuniculi*	Scabs and/or loss of fur in genital region	Penicillin var., entire herd	Do not use affected animals for breeders
Hutch burn, urine burn	Wide range of possible organisms	Similar parts affected as vent disease	Antibiotic ointment; apply to affected parts	Keep hutches clean and dry
Enterotoxemia	*Clostridium spiroforme*	Sudden death	Onset usually too rapid to treat	Avoid diets too low in fiber

INFECTIOUS DISEASE	CAUSE	SYMPTOMS	TREATMENT	PREVENTION
Mucoid enteropathy	Unknown	Constipation	None	Oral fluids to overcome dehydration
Tyzzer's disease	*Bacillus piliformis*	Severe diarrhea	None	None
PARASITIC DISEASES				
Coccidiosis: hepatic and intestinal	*Coccidia*	Young rabbits may be off feed, dull	Sulfaquinoxoline in feed or water	Good sanitation; avoid fecal contamination
Ear mites	Infestation	Scratching at ear	Clean ear with peroxide, then use ear mite medicine	Good sanitary practices
MISCELLANEOUS CONDITIONS				
Wet dewlap (moist dermatitis)	Various	Inflammation of the dewlap	Clip affected area—use antiseptic powder	Use automatic dew drop valves; elevate drinking dishes
Hair chewing	Perhaps low-fiber diet	Chewing of fur	Change to higher fiber diet	High-fiber diet, e.g., good quality hay
Hairballs	Swallowing fur	Indigestion, gagging	None	High fiber diet—mineral oil, etc. of no value
Heat Exhaustion	Overexposure to bright sunlight	Heavy panting, elevated body temperature, and extreme lassitude.	Immerse severely affected rabbits in cold water	Construct hutches to allow for good ventilation. Provide cold water and sprinkle hutches in hot weather.
Sore hocks (ulcerative pododermatitis)	Direct contact with wire mesh, especially in heavy breeds where there is an accumulation of urine-soaked feces.	Raw, weeping sores on hind legs	Difficult to treat. Affected animals should be culled.	Keep hutches clean
Malocclusion	An inherited characteristic	Failure of teeth to grind against each other, especially the incisors or front teeth.	Affected teeth should be clipped off.	Affected animals should not be used as breeders.

SHEEP

INFECTIOUS DISEASE	CAUSE	SYMPTOMS	TREATMENT	PREVENTION
METABOLIC DISEASES *Those caused by feed intake or stresses affecting all the body systems due to conditions such as lambing.*				
Grass staggers, grass tetany, hypomagnesemic tetany	Low intake of magnesium; suddenly placing animals on lush pasture in the spring	Erratic behavior after being put on lush pasture	None	Magnesium supplements to fertilize pastures; feed dry hay before turning out
Parturient paresis	Unknown	Sudden onset 6–10 weeks prior to lambing, especially in heavy ewes: muscle tremors; stilted gait; down; death	Calcium solution I.V. or under skin	Good dietary management
Photosensitization	Light colored skin that has a hyperactive reaction to sunlight	Reddening and fluid (edema) in the skin.	Graze at night; steroids	Affected animals should not be used as breeders
Some breeds of sheep (Southdowns and Corriedales) have a defect existing in liver metabolism that prevents substances that predispose the animal to photosensitization from being eliminated in the usual way. Sheep of these breeds showing this trait should not be used as breeders.				
White muscle disease	Selenium deficiency	Stiff gait, arched back unable to rise	Selenium injections	Selenium supplement in feed
Pregnancy toxemia, ovine ketosis	Any factor that disrupts feed intake in fat ewes carrying twins late in pregnancy	Nervous signs, inability to get on their feet; coma	Propylene glycol, steroids induce abortions, glucose	Induce exercise; feed away from shelter; avoid stressful situations where possible
INFECTIOUS BACTERIAL DISEASES				
Tetanus	Bacteria: *Clostridium tetani*	Often follows surgical procedures in early life; "saw-horse" appearance.	None	Administer tetanus antitoxin; later give toxoid
Overeating disease	*Clostridium* types C, D	Sudden death, especially in weaned lambs on full feed	No effective treatment	Bacterin to those on full feed

INFECTIOUS DISEASE	CAUSE	SYMPTOMS	TREATMENT	PREVENTION
VIRAL DISEASES				
Bluetongue	Bluetongue virus spread by biting flies	Difficult breathing; reddening of muzzle, lips, ears; ulcer, erosions of mouth	None	Vaccine in affected areas only
OPP (ovine progressive pneumonia)	Retrovirus	Very slowly progressing disease in mature animals; slow, wasting disease	None	Test and slaughter program; isolate lambs at birth, and feed milk from known-negative ewes
Sore mouth, contagious ecthyma	Poxvirus	Lesions on lips, sometimes feet.	None	Recovered sheep highly resistant; vaccine available
Scrapie (Transmissible spongiform encephalopathy)	Prion (structure smaller than a virus)	Loss of wool, intense itching, altered gait; loss of condition in sheep usually over 2 years	None	Test and slaughter
"Black disease" (infectious necrotic hepatitis)	*Clostridium novyi*	Sudden death; toxin produced by organism, together with liver flukes, which have already damaged the liver.	None	Toxoid (vaccine) in fluke region
Mastitis	Often *Streptococcus* or *Staphylococcus* species	Hot, swollen udder; Abnormal secretion	None	Antibiotics; hot packs; frequent stripping
Strawberry foot rot	Fungus (actinomycete): *Dermatophilus congalensis*	Sores in lower leg	Penicillin variants	Self-limiting
Scours	*Escherichia coli*	Profuse diarrhea in newborn	Fluids to combat dehydration; antitoxemia drugs, antibiotics	Dip navels in iodine; strict sanitation in lambing pens
Joint ill	*Erysipelothis rhusiopathiae*	Swollen joints and sometimes navel	Penicillin variants	Dip navels, strict cleanliness when castrating or docking
PARASITIC DISEASES *There are a wide range of parasites that affect sheep. Some invade various parts of the digestive tract while lungworms settle in the lungs. (These are largely covered in the text.)*				

GOATS

INFECTIOUS DISEASE	CAUSE	SYMPTOMS	TREATMENT	PREVENTION
CAE (caprine arthritis and encephalitis)	Lentivirus	Encephalitis in young goats, arthritis in adults	None	Butazolidin in arthritic form; remove kids from doe immediately after birth and feed pasteurized colostrum
Abscesses	*Corynebacterium pseudotuberculosi*	Abscess formation around head and neck.	None	
Urinary calculi (esp. in pet wethers)	Metabolic imbalance	Urinate with difficulty or not at all	Surgery to create artificial opening	Feed diet with calcium-phosphorus ratio of 2:1; add ammonium chloride to diets; keep magnesium level low
Mastitis	Several different bacteria	Abnormal milk ranging from watery to bloody	Variety of antibiotic preparations infused into udder; may need to run antibiotic sensitivity test to determine the best choice	Strict sanitation during the milking process; clean bedding and in general clean environment

Goats are subject to much the same range of diseases as are sheep. The management of the various conditions follows the lines followed in sheep.

PIGS

INFECTIOUS DISEASE	CAUSE	SYMPTOMS	TREATMENT	PREVENTION
Metabolic and nutritional iron deficiency	Milk naturally deficient in iron	Symptoms of anemia	None	Administer iron to newborn by injection; keep sod in pen
Hypoglycemia	Any factor that limits piglets' milk intake, especially during first week of life	Low body temperature; listless; unresponsive	Give glucose intraperitoneally	Be sure sow has enough milk; that all piglets are getting their share; keep warm and dry

INFECTIOUS DISEASE	CAUSE	SYMPTOMS	TREATMENT	PREVENTION
MMA or Mastitis-metritus-agalactia syndrome (lactation failure in sows)	Not understood at this time, a complex interaction of many factors	Piglets show signs of starvation; piglets depressed; increased temperatures; sows listless; refuse to let piglets nurse	Corticosteroids and broad-spectrum antibiotics	Good management practices; place piglets on foster sow if available
BACTERIA, SPIROCHETES, AND VIRUSES				
Bacterial brucellosis (a disease transmissible to man, especially when handling infected carcasses at slaughter)	*Brucella suis*	Abortion: temporary or permanent sterility: lameness	None	Test and slaughter
Atrophic rhinitis	A complex disease with many contributing factors	Twisted snout may be end result	Several drugs effective including trimethoprim, tylocin, and tetracycline	Keeping a closed herd; good sanitation
Enteric colibacillosis	*Escherichia coli*	Profuse watery diarrhea in young pigs	Restore fluid level; antibiotics as determined by sensitivity, previous experience	Avoid chilling and dampness; vaccinations of sows to increase level of necessary antibodies
Enteritis	*Clostridium perfringens*, Type C	Bloody diarrhea in piglets, 1–5 days	None	Vaccination of pregnant sows is of some value to increase value of colostrum in affording protection
Edema	*Escherichia coli*	Disease of young pigs 5–14 days after weaning; rapid death; wobbly; paddling of legs; head twist	Antibiotics may help	Gradual change of feed from creep feed to weaning ration

INFECTIOUS DISEASE	CAUSE	SYMPTOMS	TREATMENT	PREVENTION
Erysipelas	*Erysipelothrix rhusipathiae*	Several different forms of this disease: (1) sudden death; high temperature, walking stiffly or remain lying down; (2) may show skin discolorations as "diamond skin disease"; (3) arthritis; (4) heart valve involvement	Penicillin variants	Vaccination
Leptospirosis (may act as reservoir to cause infection in man)	*Leptospira pomona*	Abortions; being off feed; listlessness	None	Vaccinations
Mycoplasma pneumonia	*Mycoplasma hyopneumoniae*	Coughing, mild pneumonia; greatest problems: poor feed utilization and growth rate	Tylocin, Tetracycline	Establishment of disease-free herds
Pleuropneumonia	*Haemophilus pleuropneumoniae*	May be very severe with sudden deaths, especially in young pigs; extremely difficult breathing; high fatality rate	Several antibiotics have been suggested, including Tylocin and Trimethoprim	All-in, all-out management (all pigs in a group introduced simultaneously into area where they will be raised and later removed simultaneously).
Salmonellosis	*Salmonella cholerasuis*	Nursing pigs have diarrhea but die from septicemia; older pigs have bloody diarrhea	Some antibiotics may be of value, but no drastic changes may be seen	Intensive clean-up after outbreaks
Swine dysentery	*Treponema hyodysenteriae* (a spirochete)	A mucous, bloody diarrhea is the most common symptom	Do antibiotic sensitivity test and determine best antibiotic, which might include bacitracin, lincomycin, and others	Good sanitation; treatment of carrier pigs
Tuberculosis (a disease transmissable from animal to man)	*Mycobacterium* species. Pigs are susceptible to three types: cattle, bird, and man	Lymph gland involvement: a wasting disease	None	Good management (e.g. cleanliness, good ventilation)

INFECTIOUS DISEASE	CAUSE	SYMPTOMS	TREATMENT	PREVENTION
VIRUSES				
Hog cholera (no longer present in USA), now more properly called "classical swine fever" (CSF).	DNA virus	Lethargy; off feed; high temperature	Hyperimmune serum	Vaccinations not always effective; worldwide test-and-slaughter policy in effect for any animals that test positive.
Pseudorabies, "mad itch"	Herpes virus. The pig acts as a reservoir for the virus, which is highly fatal in other species such as cattle and sheep	In piglets, may see fever, trembling, convulsions; symptoms become progressively less severe as pig grows older	None	Bring in replacements from disease-free herds; segregation; isolation
Swine influenza, hog flu	Type A influenza virus	Rapidly spread throughout entire herd; high temperature; off feed; coughing; prostration	None	Good management with a stress-free environment
TGE (transmissible gastroenteritis)	Coronavirus	Vomiting followed by profuse, watery diarrhea; highly fatal in very young pigs, much less so in older pigs	None	Vaccinations
PARASITES *Those worms affecting the stomach and intestinal tract of pigs can be controlled by the same types of medications used in other species. There are some worms peculiar to swine that should be mentioned.*				
Kidney worm infestation	*Stephanurus dentatus*	Pigs tend to be "poor doers"; economic loss from contamination of affected organs and tissues	None	Maintain clean envirronment; rotate stock to keep herd young through "gilts-only" breeding program.
Trichinosis	*Trichanella spiralis*	Primarily a public health problem where insufficiently cooked, infected meat is eaten	None	Garbage containing pork should be cooked at 212° F for 30 minutes before being fed to swine, or not feed at all

CALVES

INFECTIOUS DISEASE	CAUSE	SYMPTOMS	TREATMENT	PREVENTION
Colibacillosis	*Escherichia coli*	Diarrhea; dehydration; weakness; death	Electrolytes, fluids, antibiotics	Buy calves from reputable sources; avoid stress; keep isolated; draft-free, well ventilated housing (calf hutch excellent)
Pneumonia	Various bacteria and viruses	Cough; difficulty breathing; off feed; fever	Fluids, antibiotics	

A Homemade Incubator

While you can buy an incubator for hatching eggs, a homemade one is so simple to make that it's silly not to go that route. The only thing you have to purchase is a thermostat and possibly three lamp sockets and a few feet of wire.

CONSTRUCTION

Build a wooden box, either of plywood or scrap lumber, of the dimensions indicated. The depth of the box is twelve inches. Drill a ½-inch vent hole at a height of 1 inch above the egg drawer in each of the four sides (including the door) and two holes in the top. The egg drawer should be built as indicated and slid into the incubator between the supports on each side, so that it can be pulled in or out to check the eggs and turn them.

The window in front will enable those interested to watch the hatching process without constantly opening the door, thus lowering the inside temperature. The square piece of glass can be taped on over the hole, or glued on using a good epoxy. Hinge the door and use a hook and eye or other fastener to ensure that the door closes securely. The three light sockets should be mounted about 7 inches from the floor of the incubator and wired into the thermostat. Although I have not tried it with an incubator of this design, you might try wiring the center socket so that the light is constantly on. Then you will be able

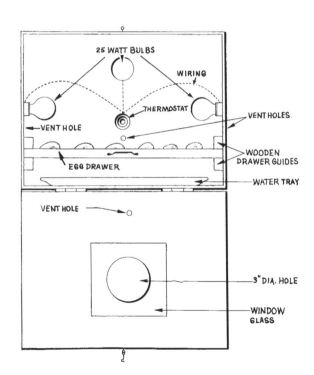

to watch the hatching without the thermostat cutting the lights off at the most exciting times.

The one bulb left on all the time should not keep the incubator too hot. If it does, replace the 25-watt bulb with a 15-watt bulb, and the incubator should operate within the desired temperature range.

The thermostat should be installed 2 to 3 inches from the level of the eggs in the egg drawer. It will then most accurately reflect the temperature of the eggs. A good unit for use in a homemade incubator can be purchased through a farm supply catalog or local hardware or feed store. The vent holes drilled into the sides and top of the incubator will ensure proper ventilation.

Table C.1: Average Hatching Period for Common Fowl	
Chickens	21 days

Geese
   Canada and Egyptian
   All other breeds
	
	35 days
	29–31 days

Ducks
   Muscovy
   All other breeds
	
	35 days
	28 days

Turkeys	28 days

OPERATION

Turn your incubator on the day before you are ready to begin hatching the eggs so that it will heat up and set the thermostat at the proper temperature. Many thermostats don't have any calibrations, and you'll have to find the correct temperature set by trial and error. You'll want to set the temperature within 99 to 103 degrees Fahrenheit. The temperature should remain between these extremes. Warm your eggs up to room temperature before placing them in the incubator so the temperature will not fall sharply

when they are put in. (See Table C.1 for average times for hatching poultry eggs.)

The pan of water will help to maintain the proper humidity in the incubator. Keep it full at all times and do not permit the water to get stagnant. Finally, turn the eggs four times a day, at 6-hour intervals if possible. Make a small mark on one side of each egg to avoid confusion when turning.

Tanning Skins

It is an unpardonable waste to throw the skins of your animals into the trash. If that is not enough, perhaps you will be convinced by the fact that you can make lovely and useful hats, gloves, and blankets from your tanned hides, and even sell and barter them for additional income or goods. Sheepskins and rabbit skins come first to mind, but don't neglect calf skins or even your goat hides. The industrious person will find uses for all these skins.

Rabbit skins are perhaps the easiest to tan because of their size, and because there is relatively little fat. Sheepskins are by far the most difficult (but you get the best finished product) because of their size and because of the large amount of fat that must be removed from the skin. Many taxidermists won't even consider tanning a sheepskin because of all the bother it can be. But that's because they scrape the fat off by hand, and that is a monumental job, believe me, because I've spent a day or two at it. It's a long and hard job and does not give very satisfactory results. The following methods will enable you to tan any type of skin (including sheepskins) with a minimum of time and effort.

PREPARATION OF SKINS

Use only skins from freshly butchered animals. Those that have been dried or salted can be used as long as they are soaked well and are pliable before tanning. Wash skins in lukewarm water with a little detergent to remove loose dirt and bloodstains. Rinse well in cold water and wash as often as is necessary to remove all blood.

FLESHING

This is why tanning has never caught on—fleshing can be terrible. With calf skins, rabbit skins, and other animal skins that don't have too much fat, you can flesh by hand in a few minutes and go right on to the next step, tanning. With sheep and goat skins, fleshing, or the removal of meat and fat from the skin, is a substantial job without the use of the pickle bath mentioned below. The fleshing/tanning bath shown here is supplied by Jerome Belanger in his book *The Homesteader's Handbook to Raising Small Livestock*.

Mix enough of the solution to cover the skin, and place it and the skin in a large stoneware crock or plastic container (don't put any acid in a metal container, or it will corrode it). Weigh the skin down so it won't rise up out of the solution. A small hide may be done in less than a week, while a large sheepskin may take up to two weeks. You will learn how long with experience, but don't worry about damaging the pelts by keeping them in too long. They can stay in almost indefinitely as long as they are stirred in the solution periodically.

If the container is small and the skins are tightly packed, take them out every day and rearrange them so all sides get exposed to the tanning solution. In a larger container, or with smaller skins, take them out every three days, stir the mixture, and re-immerse them. The temperature of the solution is quite critical—keep it as close to 70 degrees Fahrenheit as possible. I tanned some skins one fall, in our upstairs. The temperature fell into the 50s up there for a week, and the skins spoiled. Apparently it was too cool for the tanning to take place, but unfortunately not cool enough to retard the rotting process. The odor is minimal with this process, so you can keep it in your living area if the temperature elsewhere is too cool.

When finished, the hide should be taken from the solution and rinsed in cold water to remove any remnants of the acid. If all goes well, the flesh and fat will peel off the hide in large sheets. If you have ever fleshed a hide by hand, you will be amazed at how much easier this process is. What will take hours upon hours by hand will take perhaps 10 or 15 minutes after being in the acid bath. If the fleshing is still hard to complete, place the hide back in the solution for a few days until the fat peels off easily.

Once the fleshing is complete, rinse, and put back into the solution for another week. After this, wash in warm water and rinse well. I have had good results using this method alone, but if your results are not satisfactory (in regard to the quality of the leather and hide after tanning), you can go on to the tanning procedures, listed next. Again, I have found these steps unnecessary, so if you try it and agree, skip the next section and go right on to finishing.

TANNING

After fleshing, either by hand or with the above process, you can tan using one of the following methods (see Alum Mixture and Tannit Mixture boxes):

Mix with enough water to cover skins.

Dissolve the alum first in hot water, then add salt and enough cold water to achieve the correct solution. Keep the skin in this solution for 2 days and turn three times a day.

First use an acid bath as explained above. Then keep hide in Tannit solution for 5 days and stir three times a day.

Alum, an astringent, and Tannit, a buffered aluminum sulfate compound, can be purchased from many taxidermy supply companies, such as Van Dyke's (www.vandykestaxidermy.com) or Jim Allred Taxidermy Supply (www.jimallred.com).

FINISHING

If you want to remove the hair from your hide, you can pull it out after it's been either in the acid bath or in one of the tanning solutions (the hide must still be wet for the hair to come out easily).

After removing the hide from whichever solutions you use, rinse the skin in cold water to remove any solution that may be left. Then let the skin drip-

dry out of sunlight. Before the skin is completely dry and while it is still pliable, the leather side may be oiled. Then you must work the leather to make the hard, brown skin soft and white. Constant stretching, pulling, and rubbing over a smooth board will do the trick, in time. Be sure to do this work before the skin dries out. Another easy way to finish the hide is to first stretch it and nail it to a frame (any holes or cuts in the hide should be sewn up before stretching to prevent further ripping of the hide), then sand it with an electric sander, or by hand. After the leather is softened, comb burrs and other chaff out of the fleece with a metal comb, trim the skin to even it up (I usually cut the legs off a hide), and you're all set.

TANNIT MIXTURE

1 lb. Tannit AGS
2 ½ lbs. (3 ¾ cups) salt
10 gallons water

Manure Table

ANIMAL	WEIGHT (LBS.)[1]	MANURE[2] PER ANIMAL PER DAY[3] (LBS.)	PER ANIMAL PER YEAR (LBS.)[3]	% URINE	% FECES	% NITROGEN	% PHOSPHORUS	% POTASSIUM
Chicken	5	.26	95	na.	na.	1.4	0.5	0.6
Sheep and Goat	100	4.0	1,460	50	50	1.1	0.2	0.8
Pig	500	32.5	11,860	45	55	0.7	0.2	0.5
Rabbit	4	.09	34	50	50	2.4	1.4	0.6
Calf	100	6.0	2,190	50	50	0.6	0.2	0.4
Beef cow	1,000	65–80	23,725–29,200	na.	na.	1.94	0.42	1.44

[1] This is an average figure. For chickens, sheep and goats, and pigs this is an average for an adult. For the rest of the animals, it is a point between birth weight and slaughtering weight.

[2] Raw manure. That is, urine and feces with no bedding.

[3] Based on average weight from Column 1.

Note: There can be up to a 20 percent variation in the amount of manure, depending on feed, environment, and health factors. Urine, pound for pound, is of greater fertilizing value than feces (except in pigs), so good, absorbent bedding is most important. Storage is also important. Manure that is left outside will have less value than that which is under some cover.

Remember also that much of the manure will be voided on pasture. To determine how much you can save, figure the number of days the animal is confined and multiply times the entry under pounds of manure per day. Similarly, for those animals that don't live a full year, figure the number of days alive and multiply times pounds per day.

BOOKS AND OTHER PUBLICATIONS

Note: Some out-of-print books and publications have been included in this bibliography, as they remain useful references for readers looking for additional information on particular topics. Some may be found in your local library. The vocational agriculture department of your local high school or technical school might have some of these publications. For readers located near a state university with a college of agriculture, the college library would be a good source of both in-print and out-of-print publications.

Most publications of state extension services and the U.S. Department of Agriculture (USDA) are available free, or at minimal cost, from your local office of the Cooperative Extension Service. Consult your telephone directory for the office nearest you. Also, the USDA Web site at www.usda.gov is a valuable resource. For further information on this type of publication, see "Other Sources of Information" following this section.

General

Battaglia, R. A., and V. B. Mayrose. *Handbook of Livestock Management Techniques.* Minneapolis, MN: Burgess, CEPCO Division, 1981.

Belanger, Jerome. *The Homesteader's Handbook to Raising Small Livestock.* Emmaus, PA: Rodale Press, 1974, out of print.

Bundy, Clarence R., et al. *Livestock and Poultry Production.* Englewood Cliffs, NJ: Prentice-Hall, 1982.

Ensminger, M. E., and R. O Parker. *Sheep and Goat Science* (6th ed.). New York: Prentice Hall, 2001.

Haynes, N. Bruce, DVM. *Keeping Livestock Healthy: A Veterinary Guide to Horses, Cattle, Pigs, Goats & Sheep* (4th ed.). No. Adams, MA: Storey Publishing, LLC, 2001.

Iowa State University. *Livestock Waste Facilities Handbook* (2nd ed.). Ames: Iowa State University, Midwest Planning Service, 1985.

Merck and Co., Inc. *The Merck Veterinary Manual.* Rahway, NJ: Merck and Co., 1991.

Wilson, J. V. *Weaning Is for Anyone.* New York: Van Nostrand Reinhold, 1975, out of print.

Ziegler, P. T. *The Meat We Eat* (12th ed.). Danville, IL: Interstate Printers and Publishers, 1985.

Poultry

Banks, Stuart. *The Complete Handbook of Poultry Keeping.* New York: Van Nostrand Reinhold, 1979, out of print.

Graves, Will. *Raising Poultry Successfully.* Charlotte, VT: Williamson, 1985.

Mercia, Leonard S. *Storey's Guide to Raising Poultry*. (revised). No. Adams, MA: Storey Publishing, LLC, 2000.

Robinson, Ed. *Producing Eggs and Chickens with the Minimum of Purchased Feed*. Charlotte, VT: Garden Way, 1972.

Rabbits

Bennett, Bob. *Storey's Guide to Raising Rabbits*. No. Adams, MA: Storey Publishing, LLC, revised, 2000.

Bennett, Bob. *Raising Rabbits Successfully*. Charlotte, VT: Williamson, 1984.

Cheeke, Peter R. *Rabbit Feeding and Nutrition*. Orlando, FL: Academic Press, 1987.

Cheeke, P. R., N. M. Patton, S. D. Lukefahr, and J. I. McNitt. *Rabbit Production*. (6th ed.) Danville, IL: Interstate Printers and Publishers, 1987.

Choate, Peter. *Rabbit Production*. Charlotte, VT: Williamson, 1987.

Kanable, Ann. *Raising Rabbits*. Emmaus, PA: Rodale Press, 1980.

Sanford, J. C. *The Domestic Rabbit*. Somerset, NJ: Wiley, 1974, out of print.

Templeton, G. S. *Domestic Rabbit Production*. New York: Scribner, 1975.

Thear, Katie. *Practical Rabbit-Keeping*. Woodstock, NY: Beekman, 1988.

Sheep

Juergenson, Elwood M. *Approved Practices in Sheep Production*. (4th ed.) Danville, IL: Interstate Printers and Publishers, 1981.

Kruesi, William K. *The Sheep Raiser's Manual*. Charlotte, VT: Williamson, 1985.

Midwest Plan Service. *Sheep Housing and Equipment Handbook*. Ames, IA: Midwest Plan Service, 1994.

Sheep Industry Development Program, Inc. (American Sheep Industry Association). *The Sheep Production Handbook*. (revised). Centennial, CO: ADS/Nightwing Publishing, 2003.

Simmons, Paula. *Storey's Guide to Raising Sheep: Breeds, Care, Facilities*. No. Adams, MA: Storey Publishing, LLC, 2000.

Goats

Belanger, Jerry. *Storey's Guide to Raising Dairy Goats: Breeds, Care, Dairying*. No. Adams, MA: Storey Publishing, LLC, 2000.

Halliday, Jill and John. *Practical Goatkeeping*. Woodstock, NY: Beekman, 1990.

Luttman, Gail, *Raising Milk Goats Successfully*. Charlotte, VT: Williamson, 1986.

Pigs

Belanger, Jerome. *Raising the Homestead Pig*. Emmaus, PA: Rodale Press, 1977, out of print.

Kellogg, Kathy and Bob. *Raising Pigs Successfully*. Charlotte, VT: Williamson, 1985.

Klober, Kelly. *A Guide to Raising Pigs: Care, Facilities, Breed Selection, Management*. No. Adams, MA: Storey Publishing, LLC, 1997.

Krider, J. L., J. H. Conrad, and W. E. Carroll. *Swine Production*. (5th ed.). Heightstown, NJ: McGraw-Hill, 1982, out of print.

Pond, W. G., and J. H. Maner. *Swine Production and Nutrition*. New York: Van Nostrand Reinhold, 1984.

van Loon, Dirk, *Small-Scale Pig Raising*. No. Adams, MA: Storey Publishing, LLC, 1978.

Veal

Popow, Jeffrey S., ed. *Special-fed Veal Production Guide*. Ithaca, NY: Northeast Regional Agricultural Engineering Service/Cooperative Extension, 1991.

Guide for the Care and Production of Veal Calves, (4th ed.). North Manchester, Indiana: American Veal Association, Inc., 1993.

Beef

Grohman, Joann S. *Keeping a Family Cow*. (5th rev. ed.). Dixfield, ME: Coburn Farm Press, 2000.

Thomas, H. S. Storey's *Guide to Raising Beef Cattle: Health/Handling/Breeding* No. Adams, MA: Storey Publishing, LLC, 2000.

Hobson, Phyllis. *Raising a Calf for Beef*. No. Adams, MA: Storey Publishing, LLC, 1976.

van Loon, Dirk, *The Family Cow*. No. Adams, MA: Storey Publishing, LLC, 1976.

Feeds and Feed Crops

Perry, T. W., A. E. Cullison, and R. S. Lawrey. *Feeds and Feeding*. (6th ed.) New York: Prentice Hall, 2002.

Langer, Richard, Susan McNeill. *Grow it! The Beginner's Complete In-Harmony-With-Nature Small Farm Guide—From Vegetable and Grain Growing to Livestock Care*. New York: Noonday Press, 1994.

Miller, D. A. *Forage Crops*. Heightstown, NJ: McGraw-Hill, 1984.

Morrison, Frank B. *Feeds and Feeding* (23rd ed.) Ithaca, NY: Morrison Publishing Co., 1967. (Although out of print, this text has long been considered the bible of books on feeds and feeding.)

Bradley, F. M., B. W. Ellis (Eds.). *Rodale's All-New Encyclopedia of Organic Gardening*. (reprint) Emmaus, PA: Rodale Press, 1993.

OTHER SOURCES OF INFORMATION

Your local county extension agent will offer free advice on raising livestock and growing crops. Many excellent publications, some of them free, are available through the Extension Service. Write to the Extension Service Director at the College of Agriculture at your state university, or check the extension service Web site at www.csrees.usda .gov.

On the national level, you can obtain additional information from:

Government Printing Office
Superintendent of Documents
732 North Capitol St. NW
Washington, DC 20401
202-512-0000
Web site: www.gpo.gov

Federal Consumer Information Center
Dept. WWW
Pueblo, CO 81009
Online catalog, or paper copy can be ordered online
1-888-878-3256
Web site: www.pueblo.gsa.gov

USDA Economic Research Service
1800 M Street NW
Washington, DC 20036-5831
1-800-999-6779
Web site: www.ers.usda.gov/publications/

The USDA (United States Department of Agriculture) offers many publications—most of them free, and many available online—on raising livestock and crops, such as the USDA Yearbooks and the *Agriculture Fact Book*, all of which are published annually, as well as numerous booklets and supplemental pamphlets. Hard copies are available for sale from the Government Printing Office (see address above). Yearbooks are about $28.00 and the *Agriculture Fact Book* is currently $26.00 per copy. Issues from previous years are also available. These can be ordered by phone, by mail, or online at bookstore.gpo.gov, or can be downloaded free online. Most pamphlets and supplements are also available online.

PERIODICALS

American Farmland
American Farmland Trust
1920 North Street NW, Suite 400
Washington, DC 20036
$25 dues for annual membership; includes quarterly subscription.
This is a non-profit organization dedicated to the preservation of American farmland.
Web site: www.farmland.org

Renewable Agriculture and Food Systems
Cambridge University Press
1 Liberty Plaza
New York, NY 10006
newyork@cambridge.com

Farm Journal
Farm Journal, Inc.
230 West Washington Square
Philadelphia, PA 19105
1-800-331-9310
$24.75 for 12 issues per year.
For owners and operaters of family farms and ranches.
Web site: www.agweb.com/farmjournal.asp

Countryside and Small Stock Journal, Backyard Poultry, sheep! magazine, and Dairy Goat Journal
Countryside and Small Stock Journal
500 Mallory Way
Carson City, NV 89701
1-800-551-5691
$18 per year for *Countryside;* all others, $21 per year; 6 issues (all publications).
Web site: www.countrysidemag.com/

Progressive Farmer
Box 2581
Birmingham, AL 35202
$12 per year.
Published monthly.
Web site: www.progressivefarmer.com/

Successful Farming
Meredith Corp.
1716 Locust Street
Des Moines, IA 50336
$15.95 per year.
Published monthly.
Order online at Web site: www.LHJ.com/

Poultry Press
PO Box 542
Connersville, IN 47331
765-827-0932
$23 per year.
Caters to poultry breeders, but the articles seem
    well-written and pertinent to anyone involved
    in livestock; has a very down-home, family
    approach.
e-mail: info@poultrypress.com
Web site:www.poultrypress.com

Wallaces Farmer
Farm Progress Companies, Inc.
6200 Aurora Ave # 609E,
Urbandale, IA 50322
$20 per year outside Iowa; $12 per year for Iowa res-

idents. Subscriber must demonstrate employment
in occupation related directly to agriculture.

Small Farmer's Journal
Small Farmer's Journal, Inc.
192 W. Barclay Drive
PO Box 1627
Sisters, OR 97759-1627
1-800-876-2893
$35 per year.
Published quarterly.
Web site: www.smallfarmersjournal.com
E-mail: agrarian@smallfarmersjournal.com

Lancaster Farming
PO Box 609
Ephrata, PA 17522
$43.00 per year to nearby states, $56 to other areas;
    check Web site for more information.
Published weekly.
Web site: www.lancasterfarming.com

The Shepherd
Sheep and Farm Life, Inc.
PO Box 168
Farson, WY 82932
$14 per year.
Published monthly.
E-mail: sheepmag@bright.net

Periodicals are also published by many of the various colleges of agriculture at state universities.

Check with any state university and also search their Web site (not just in your own home state!) to see what is available from them, by mail or online.

Some excellent farm magazines are circulated on a regional basis. Check with your county extension office or the Extension Service Web site (www.csrees.usda.gov) for information regarding such publications in your area.

ONLINE RESOURCES

American Small Farm, online publication; there is a mailed, hard-copy magazine available as well; Web site: www.smallfarm.com

Western Livestock Journal, "The National Livestock Weekly Online"; Web site: www.wlj.net

Drovers, free-to-the-industry online trade publication pertaining to the beef industry; e-mailed weekly; Web site: www.drovers.com

Watt's Pig International, free to industry professionals; covers breeding, care, feeding, marketing, and more; Watt Publishing is located in Illinois and has other trade publications, also available free; Web site: I.NL02.net/watt0007/

Virtual Livestock Library contains links to, well . . . everything; Web site: www.ansi.okstate.edu/library/

Livestock Research for Rural Development, published by Fundación CIPAV, Cali, Colombia; an international online journal researching sustainable agriculture in the developing world; Web site: www.cipav.org.co/lrrd/lrrdhome.html

The Livestock Conservancy is a nonprofit organization devoted to protecting the perpetuation and the genetic diversity of rare breeds of livestock, through education and research, and as a means of sustainable agriculture for both the small and large farmer; also provides updates on disease threats and concerns, and acts as an advisory organization to the USDA; Web site: www.livestockconservatory.org

SOURCES OF SUPPLIES

Here is a list of sources of supplies, from farm machinery to day-old chicks. All provide catalogs; some companies may charge for them. Many also have Web sites, which list their products for sale and additional contact information.

Forestry Suppliers, Inc.
205 W. Ranlain St.
PO Box 8397
Jackson, MS 39204
Forestry, farming equipment.
Free catalog.
Web site: www.forestry-suppliers.com

Midstates Wool Growers Cooperative Association
9449 Basil-Western Rd.
Canel Winchester, OH 43110
Livestock supplies.
Free catalog.
Web site: www.midstateswoolgrowers.com

A.M. Leonard
6665 Spiker Road
Piqua, OH 45356
Gardening and farming equipment.
Free catalog.

NASCO
901 Jamesville Ave.
Ft. Atkinson, WI 53538
Farm and ranch supplies.
Free catalog.
1-800-558-9595

Murray McMurray Hatchery
Box 458, 191 Closz Drive
Webster City, IA 50595-0458
Chickens.
Free catalog.
Web site: www.mcmurrayhatchery.com

Stromberg's
PO Box 400
Pine River, MM 56474
Chickens, poultry equipment.
Free catalog.
Web site: www.strombergschickens.com

Country Hatchery
PO Box 747
Wewoka, OK 74884
Live poultry.
Free catalog.
405-257-8315

Grain Belt Hatchery & Poultry Farm
PO Box 125
Windsor, MO 65360
Exotic as well as the more popular breeds of
    chickens.
50 cents for catalog.
e-mail: gpmr@iland.net

Reich Poultry Farms, Inc.
1625 River Road
Box 14K
Marietta, PA 17547
Chickens.
Free catalog.
717-426-3411

The Poultry Connection Web site is also an excellent
source for numerous other suppliers. Web site: www
.poultryconnection.com

Balling gun. An instrument that is slipped partly down the throat of an animal and used to administer large pills.

Barrow. A male pig castrated before sexual maturity.

Boar. A male pig.

Bolus gun. See Balling gun.

Broiler. A chicken of either sex that weighs 2½ pounds and is less than 8 months old (cf. Fryer and Roaster).

Broken mouth. A condition occurring in sheep and goats, usually at about the age of 5 to 6 years, whereby some of the permanent teeth are missing from the mouth (cf. Full mouth, Gummer).

Brooder. The enclosure in which young animals are kept, or are allowed free access to, which furnishes heat until they adapt to outside temperatures.

Broodiness. A natural condition in poultry, in which a female goes out of production and will then set on and attempt to hatch eggs.

Broody coop. An enclosure used to confine a broody hen and hasten her return to normal production.

Buck. A male goat. A male rabbit. Less commonly: a male sheep.

Capon. A male chicken castrated surgically before sexual maturity (cf. Stag and Caponette). In butchering, a castrated male chicken that weighs 6 to 8 pounds.

Caponette. A male chicken that is neutered before sexual maturity by the implantation of female hormones (cf. Capon).

Carbonaceous hay. See Grass hay.

Castrate. To remove the testicles of a male animal.

Chaff. Unwanted parts of grain separated during winnowing.

Chevon. Goat meat.

Chick. A young chicken.

Closed-faced. A sheep that has considerable wool covering about the face and eyes. This often leads to a condition known as wool blindness (cf. Open-faced; also Wool blindness).

Cock. Any male chicken butchered after 8 months of age. A stag.

Cockerel. A male chicken less than 1 year old.

Colostrum. The milk produced by an animal immediately and for the first few days after the birth of its young. Rich in antibodies that protect the animal against many diseases. A young animal that does not receive this milk may die or will be very difficult to raise.

Creep feeding. The feeding of a young animal by means of an enclosure accessible to it, but not to its mother.

Crop. A digestive organ in poultry in which food is prepared for digestion.

Crossbred. An offspring that results from the breeding of two purebred parents of different breeds. A hybrid.

Crossbreeding. The mating of purebred parents from different breeds (cf. Hybrid vigor).

Cull. (v.) To remove an inferior animal from a flock. (n.) Any animal that is culled.

Dicalcium phosphate. A mineral mix rich in calcium and phosphorus.

Disbud. To remove the horns of an animal.

Dock. (v.) To cut short the tail of an animal (most commonly a lamb); usually for sanitary reasons and to facilitate breeding in females. (n.) The area around the tail of sheep or other animals.

Doe. A female goat. A female rabbit.

Drake. A male duck.

Dress out. To remove the feathers or skin and to cut up and trim the carcass of an animal after slaughter.

Drylot. An area of confinement containing little or no natural feed into which all or most of an animal's feed must be brought (cf. Free-ranging).

Duck. A female duck.

Duckling. A young duck of either sex.

Elastrator. A tool used in castrating and docking. A tight rubber band is applied to the tail or the scrotum; the circulation is thereby cut off and the tail or scrotum gradually dries up and falls off.

Emasculatome. A tool used for docking and castrating that both cuts and crushes surrounding tissue to prevent excessive bleeding.

Ewe. A female sheep.

Eyeing. Clipping the wool from around the face of closed-faced sheep to prevent wool blindness.

Farrow. To give birth to a litter of piglets.

Feed conversion ratio. The rate at which an animal converts feed to meat. If an animal requires 4 pounds of feed to gain 1 pound it is said to have a four to one (4:1) feed conversion ratio.

Field crops. Feed plants grown primarily for their seeds. For example, corn, wheat, oats, soybeans (cf. Forage crops).

Finishing. The increased feeding of an animal just prior to butchering, which results in rapid gains and increased carcass quality.

Flail. (n.) a tool used for threshing grain. (v.) To use a flail to thresh grain.

Flushing. The practice of increasing the feed intake of a female animal just prior to ovulation and breeding. This causes the animal to gain some weight and drop more eggs, often resulting in larger litters.

Forage crops. Those plants or parts of plants that are used for feed before maturing or developing seeds (cf. field crops). The most common forage crop is simple pasture.

Fowl. For butchering purposes, a female chicken that is more than 8 months old.

Free-choice feeding. A type of feeding routine whereby feed, water, and salt are provided in unlimited quantities and an animal is left to regulate its intake (cf. Hand feeding).

Free-ranging. Allowing animals, especially poultry, to roam freely and eat as they wish without any sort of confinement (cf. Drylot).

Freshen. To come into milk, as when a dairy animal gives birth.

Fryer. A chicken of either sex that weighs 2½ to 3 pounds and is less than 8 months old (cf. broiler and roaster).

Full mouth. A state in sheep or goats when an animal has a full set of permanent teeth. This occurs at approximately the age of 4 years. The animal will continue to have what is known as a full mouth until it loses some teeth (cf. Broken mouth) or until it loses all of its teeth (cf. Gummer).

Gander. A male goose.

Garbage. Scraps. Leftover food. A very poor choice of words, as it has bad connotations.

Gilt. A female pig that has not yet produced a litter (cf. Sow).

Girth. The measure of the distance around an animal at a point just behind the front legs.

Gizzard. The muscular enlargement of the alimentary canal of poultry that immediately follows the crop; has thick, muscular walls and a tough horny lining for the grinding of the food.

Goose. A female goose.

Gosling. A young goose of either sex.

Grade. An animal of no distinguishable breed or background (cf. Purebred).

Grading up. The practice of improving a flock whereby purebred sires are mated to grade animals and their offspring. In 3 generations the offspring will be ⅞ purebred and in some cases eligible for registration. Upgrading.

Grass. Any of the members of the family *Gramineae* (such as timothy, orchardgrass, Sudangrass, etc.). When used as pasture, hay, or in silage, they provide more energy than legumes but are lower in protein and vitamins (cf. Legume).

Grass hay. Any hay totally or primarily from a grass crop (cf. Legume hay; Grass; Legume).

Grass lamb. A lamb that is dropped in the springtime and is raised on pasture in the summer months, and is butchered in the fall when pasture has died.

Grit. Any ingested rough, undigestible matter, such as bits of glass or small stones, that is used to grind food in the gizzard of a chicken or other poultry.

Gummer. A sheep or goat having no teeth. (cf. Full mouth, Broken mouth).

Hand feeding. A type of feeding routine whereby an animal is fed measured amounts of food, water, salt, etc. (usually just food) at fixed intervals (cf. Free-choice feeding).

Hay. Any crop (most often grasses or legumes) that is cut, dried, and stored (either baled or loose) for later use.

Heifer. A cow that is under 3 years of age and has not yet produced a calf.

Hen. A female chicken. A female turkey.

Heterosis. See Hybrid vigor.

Hothouse lamb. A lamb that is dropped in the fall or early winter and is marketed at an age of 6 to 12 weeks. These are usually sold in periods before the Christmas or Easter holidays to take advantage of higher prices.

Hybrid. See crossbred.

Hybrid vigor. The increase of size, speed of growth, and vitality of a crossbreed over its parents. Heterosis.

Inbreeding. The mating of very closely related animals, such as mother to son, father to daughter, brother to sister. In experienced hands, this method can be used to selectively maintain certain desirable traits; if used improperly, it can produce undesirable traits and downgrade stock (cf. Linebreeding).

Kid. (n.) A young goat. (v.) To give birth to a young goat.

Kindle. (n.) A young rabbit. (v.) To give birth to a litter of rabbits.

Lamb. (n.) A young sheep of any sex less than 1 year old. (cf. Mutton, Yearling). (v.) To give birth to a lamb (or lambs).

Lambing loop. A length of smooth or plastic-coated wire used as an aid in difficult lambing.

Legume. Any of the members of the family *Leguminosae* (such as clovers, alfalfa, trefoil, vetches). When used as pasture, hay, or in silage, they provide more protein and vitamins than a comparable grass crop (see Grass).

Legume hay. Any hay made totally or primarily from a legume crop (cf. Carbonaceous hay; Grass hay; Legume).

Linebreeding. Very similar to inbreeding but the breeding is not so close; for example, the mating of cousins (cf. Inbreeding).

Milk replacer. A powder that, when mixed with water, is fed to young animals as the milk portion of their diet.

Molt. To shed feathers, fur, skin, or horns and replace them with new growth.

Mutton. In butchering, any sheep over 18 months of age (cf. Lamb, Yearling).

Needle teeth. The eight sharp teeth present in newborn piglets. Especially in large litters, these teeth can cause injury to other piglets and the sow's udder, and should be clipped (see text).

Open-faced. Sheep that naturally have little or no wool covering about the face and eyes. This is a desirable trait as it discourages the problem of wool blindness. (See Wool blindness; cf. Closed-faced.)

Piglet. A young pig of either sex.

Pinning. The sticking of a young lamb's tail to its anus. This will prevent normal bowel action and result in constipation and, if not loosened in time, death.

Polled. Without horns; either naturally or by artificial means.

Poult. A young turkey of either sex.

Pullet. A young hen less than one year old.

Purebred. An animal of a recognized breed whose lineage has been kept pure (i.e., not mixed with another breed) for many generations (cf. Grade).

Purebreeding. The mating of purebred parents from the same breed (cf. Crossbreeding).

Ram. An uncastrated male sheep or goat.

Replacement animal. A young animal that is being raised to take the place of an older animal that is being culled; e.g., since a person wants to keep a flock of five breeding ewes, he will have to buy two replacement ewes to take the place of the two older breeders that have died or are being culled.

Ringing. Clipping a breeding ram around the neck, belly, and penis in order to facilitate mating.

Roaster. A chicken of either sex that weighs 3-½ to 5 pounds and is less than 8 months old (cf. Broiler and Fryer).

Rooster. A male chicken.

Ruminant. Any one of a class of animals—including sheep, goats, and cows—that have multiple stomachs. They are most efficient feeders, because bacterial action in one of the stomachs, the rumen, increases the food value of low-grade food.

Scours. Technically, a bacterial infection in calves and sheep that results in a whitish-yellow, foul-smelling diarrhea. Informally, any diarrhea.

Scraps. Edible refuse that is saved from table scraps or collected from restaurants, food stores, etc. and fed to livestock.

Settle. To become pregnant.

Sex-linked. Distinguished sexually at birth, as chicks, by differences in coloration.

Shoat. See Piglet.

Silage. A feed consisting of certain roughages and/or field crops that are finely chopped, tightly packed in an airtight container, and allowed to ferment in the absence of air.

Sow. A female pig that has produced a litter (cf. Gilt).

Stag. A male chicken castrated after sexual maturity (cf. Capon). Also, any male chicken butchered after 8 months of age. A cock. A male pig castrated after sexual maturity.

Strip. To remove milk from the teat of an animal by sliding the fingers from the base of the teat to the end.

Swill. Scraps, used as feed, especially for pigs; usually mixed with water or milk.

Tagging. Clipping the wool from around the dock of a ewe so that it does not interfere during mating.

Teasing. Keeping a ram in sight of, but not in contact with, ewes just prior to breeding. This often stimulates ovulation in the ewes.

Thresh. To separate grains from the plant, as in removing oats from straw.

Tom. A male turkey.

Uterine capsule. A medication administered to a ewe to prevent infection after entering her with a hand or other instrument during lambing.

Wattles. Fleshy appendages hanging from the neck of a chicken, turkey, or goat.

Wean. To remove a young animal from its mother and accustom it to food other than its mother's milk.

Wether. A castrated male lamb; a castrated male goat.

Windrow. Hay or any other crop raked into a row to dry.

Winnow. To remove chaff from grain by using a current of air.

Wolf teeth. Needle teeth.

Wool blindness. A condition that develops most often in closed-faced sheep due to irritation of the eyes by wool and particles of chaff contained therein (see Closed-faced and Open-faced).

Yearling. Any animal aged 1 year to 18 months. (In sheep, cf. Mutton and Lamb.)